Working with the Curlew

A Farmhand's Life

Trevor Robinson

With illustrations by Jane Hassall
and a Foreword by Gus Dermody

Green Books

First published in 2003
by Green Books Ltd
Foxhole, Dartington
Totnes, Devon TQ9 6EB

Reprinted 2004

Cover design by Rick Lawrence
samskara@onetel.net.uk

Cover photo of the author by Maureen Merone

Text printed by MPG Books, Bodmin, Cornwall, UK
on Five Seasons 100% recycled paper

British Library Cataloguing in Publication Data
available on request

ISBN 1 903998 34 4

Acknowledgements

I would like to thank Margaret Thompson for reading my atrocious writing and typing the manuscript; Maureen and Malcolm Merone for photographs; Barbara Sykes of Freedom of Spirit, Walt Unsworth and Andrew Gagg for their encouragement; Jane Hassall for her drawings; and my long-suffering wife!

Foreword

by Gus Dermody

Although I was born ten years after Trevor Robinson, as boys we both had very similar interests. All my spare time was spent on a farm near home in Cheshire. The area was flat and fertile, so it was a 'mixed' farm. Horses were still the main power source, but tractors were starting to appear. Steam traction engines were being laid up in their hundreds, and I remember one agricultural contractor 'sheeting up' fourteen traction engines, most never to move again, save to a scrapyard.

On reading Trevor's book, thoughts came flooding back. His eye for detail and his memory retention is superb. It will be obvious right from the start that Trevor was no ordinary 'farmhand'. He asked too many questions, and his thirst for farming knowledge could barely be quenched. He actually starting shepherding sheep and cattle in the way I could only dream of doing—what I called 'dog and stick' farming, with no mechanical aids. Man attempting to work with nature in a very beautiful but at times very inhospitable area of the Yorkshire Dales.

It's ironic that many of the aspects of Trevor's early work are just the same today. The main changes for a shepherd are in the use of medicines and transport. Probably the biggest change is in the popularity of the ATV (All Terrain Vehicle, or 4-wheel motorcycle). The shepherd and his dogs are still there, but now they look after two or three hills because of the time saved in getting from A to B by motorbike, so where one shepherd looked after 400–600 ewes he now has 1000–1500, and sometimes more, with contractors doing some of the main jobs.

The social side is very well documented, with 'do's' in the village halls, sport, agricultural shows and village life in general which includes the pubs and churches, not forgetting ferrets and ferreting.

When Trevor mentions being caught 'tickling a trout', having to hand over the fish to the water bailiff and then be threatened with criminal court proceedings, it reminds me of my early career days. My parents persuaded me not to go into farming and I joined a county police force. Late one Saturday I was on duty and the town was quiet, so spying my chance I went the back way into a town centre pub. The landlord had just handed me a full pint when a voice boomed out behind me. It was the Duty Sergeant. I was told that disciplinary proceedings would be instigated for drinking on duty, and the licensee would be prosecuted for 'aiding and abetting', a sackable offence. For months I was in fear and never put a foot wrong—nothing happened. About five years later I found out the Sergeant confided to a colleague that catching me out was the best way of getting me to toe the line—he had no intention of taking it further. No doubt Trevor's tickled trout was cooked to perfection by the bailiff's wife, and they enjoyed their meal while the 'culprit' suffered long and hard wondering if proceedings were to be started.

It was all part of an early learning curve which Trevor describes in such detail. Those who were alive in the late 1940s will have some wonderful memories brought back to life, and those born later will have an authentic glimpse of the ups and downs of farming in the critical pre-mechanical era.

A brilliant book, telling of one man's start on the farming ladder—a book I couldn't put down until the end.

Gus Dermody is Commentator & Presenter
of BBC TV's 'One Man and His Dog'

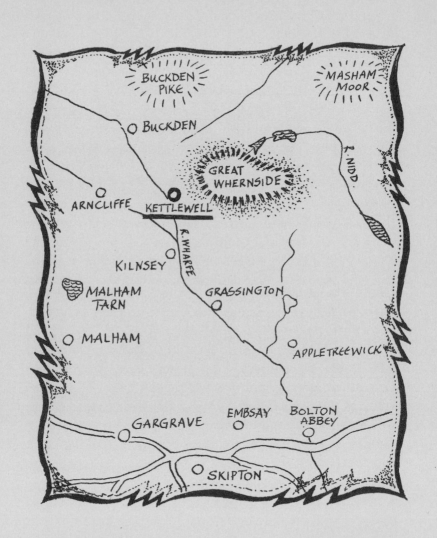

Kettlewell

I AM QUITE an old man now. I was born in Huddersfield in 1930. At eight years old, I worked before and after school on a local farm bottling milk and delivering it around Huddersfield, which was at that time a busy town full of mills producing the finest cloth in the world, Worsted. The farmer had a small herd of Ayrshire milk cows. There were many of these small farms around the West Riding towns, which delivered milk by horse and cart each morning. Bottling was done by hand, using a quart measuring ladle, and the bottles were covered with waxed cardboard tops which were also put on by hand.

Not all milk was bottled—there were also one or two milk churns in the cart, and people used to come out of their houses with milk jugs to be filled. The horse would do this round every day and it would know where to stop on every road. The outward journey got slower and slower the further they got from the farm, but once the last delivery was made the horse turned and came home at a trot.

About this time, my parents took a cottage in Kettlewell called Sunter's Cottage, right next door to Sunter's Garth Farm, owned by Mr Matt Middlemiss. This farm had a great influence on my life in the Yorkshire Dales. I first went up on to Great Whernside when I was nine years old—the two thousand acre moor was awesome, but I loved it and could not wait to leave school and work as a shepherd on it. I never had any doubt that this would happen. All holidays and most weekends I would be at Sunter's Cottage learning to be a shepherd.

WHEN I WAS OLD ENOUGH to work, I was delighted to be given a job as shepherd at Sunter's Garth by Mr Middlemiss, who was known in the village as 'Mouse'. He had about 300 acres of lowland with moor rights on Great Whernside. He was also a Moor Shepherd, which meant that he looked after all the sheep on the moor for about eight farmers in the village. The number of sheep each farmer was allowed on the moor was calculated by what are known as sheep gates. If you have ever walked in the wild country you may have seen a dry stone wall with a hole in it about 2' 6" high and about 2' wide, well made, with upright side stones and a large lintel on the top. These are sheep gates.

The farmer who owned the field with these gates was allowed to put five sheep through each one. So if there were thirty of these gates on his farm he could put 150 sheep on the moor. The system ensured that the moor didn't get over-grazed and become poor. It also meant that we knew how many sheep we had to look after, so we could count them and know when we had seen them all. We also knew how much each farmer had to pay us for caring for his sheep on the moor. I don't know how much Mr Middlemiss got, but I received 10 shillings (50p) a week and my keep. I lived in the farmhouse with Mr and Mrs Middlemiss. This was in 1946.

Our sheep on the moor were Dalesbred. It is a breed not found outside of the Dales but it did well on our moor. We had to see them as often as the weather would let us. We would be up early in the morning to see if the 'tops' were clear. If they were, Mrs Middlemiss would make a huge pile of sandwiches, mostly cheese and onion. We wore corduroy trousers and tweed jackets. Our mackintoshes were folded lengthways over one shoulder and then tied with string around the waist like a Scottish plaid. On our feet we wore Fell boots with hobnails in the soles (I still have a pair, and they weigh six pounds!). We always wore a hat or a cap. We kept the sandwiches in our pockets but later we used an old gas mask case (much better). We would let the dogs out, five of them, and with our third leg (a shepherd's crook), off we would go.

The first stop would be Hagg Dyke. This was an empty farmhouse on the edge of the moor. We would get a fire going and make a 'brew'. Leaving most of the sandwiches there, we would set off over the moor. Sheep on the mountains of Britain have painted lines on them called moor marks; they are large and can be seen from a distance. Each farm has its own mark. Some are on the shoulder, others on the middle or rump, just on one side. They are different colours and shapes. Our mark was a red stripe on the left flank. If the sheep were walking the same way as us or to our left side, we were unable to see the mark, and when they were half a mile away this could be a problem. To try and make the sheep turn around we used to whistle, as if commanding a dog. The sheep would stop and look back, then turn around to see where the dog was. We could then see the moor mark.

If this did not work, we would have to send a dog out. This had two disadvantages. First, we did not want the sheep brought to us, and secondly it would be extra mileage for the dogs in very rough country. Some days these dogs would run up to 100 miles, so we tried to save them whenever we could. We would work round about two-thirds of the moor, pushing the sheep towards the summit, and coming back to Hagg Dyke for lunch. After our sandwiches and tea we would do the last part of the moor, then head back home to Sunter's Garth. The cows would then be milked, and the milk taken around the village. We used to supply one or two houses and the youth hostel.

When on the moor we used to command the dogs by whistle. Although I could whistle very well with my fingers, I always used a home-made tin whistle because a lot of my days on the moor were very cold, and, apart from the discomfort, it was hard to get just the right tone with frozen hands. I used to make my whistle out of an old cocoa tin. I would cut a circle about two inches in diameter with an old pair of scissors from one side of the cocoa tin, then bend it in two around a two-inch nail. Next, I would punch a hole in each side with the nail, and make it smooth around the edges. Today, they are made of plastic. They are called Fortune Whistles, because they are manufactured by a company owned by Dick Fortune, who used to be one of the best sheepdog handlers in the country.

We had one dog called Spot who would take commands from my home-made whistle when he was over a mile away. Often I could not see him, but I knew where he was by the movement of the sheep.

My days shepherding on the moor, often on my own, were wonderful. I would have liked everyone to enjoy the beauty of my world. There was not only the changing of the seasons, from the curlews of spring to the howling snow-storms of winter, but also the ever-beautiful changing of the day. I remember one day in the summer setting out to do a full gathering (that is to clear all the sheep from the moor). All the farmers in the village came to help us, and it had been arranged for a long time so the weather had got to be pretty bad to stop us. It was not very good early in the morning, but we set off. The plan was that we would all start off in a semi-circle around the north side of the moor, with about half a mile between each man. When we were all in position we would move forward all together and sweep up all the sheep. As we came over the top and down the south side of Great Whernside we would close up like a purse string and finish up with all the sheep in one big flock in the collecting yards at Hagg Dyke.

It was not too bad weatherwise, and soon we were all moving steadily up towards the summit. It began to get a bit misty, however, and by the time we reached the top there was thick fog. Being Moor Shepherds, we had the farthest distance to go as we knew the moor, and with just two dogs and a small flock of sheep I felt as though I was on a different planet. As I could not see more than a few yards, the dogs had to work on their own and I just hoped that they were in contact with the men on either side of me so that no sheep could get behind us. Then, as I came down over the summit to the south side, I suddenly walked out of the cloud into bright sunlight. It was a different world; the sheep came to life, and whereas ten minutes before there had not been a sound except the odd bleat, now the whole air was filled with their calls. There to my left was the boss, and on my right another shepherd. The line was still the same shape it was before going into the cloud but now we also had a long line of sheep in front of us.

As we came lower down, the side shepherds came in and we were able to shout to one another. By the time we arrived at Hagg Dyke we were only a few yards apart with a huge mob of sheep. What with their bleating, the dogs (about twenty of them) barking and the men shouting, it was bedlam. It is quite unusual for sheepdogs to bark, but it was not often that so many were at such close quarters and in such an exciting situation, because we had to force two thousand ewes into pens that looked as though they were made to hold a thousand.

Once the sheep were all in, work really began. Most of the dogs lay on the tops of the walls and watched. We went through the flock and first pulled out Mr Lambert's, and off he went back to Kettlewell with his flock. Then we drew Mr Clough's and off he went, and so on until eight farmers had gone. That left Mr Middlemiss and me with our sheep and all the strays—the ones that had come on to our moor from the surrounding areas. It was hard work, and of course we became a man short every time a flock went. But strangely enough, it never seemed like hard work. It is a bit like lambing. I have lambed ewes for forty years, and I still get a thrill every year. Although you are so tired that you could fall asleep leaning on a fence, it is always a privilege and a pleasure to do it.

It was at one of these gatherings at Hagg Dyke that I first saw two rams fighting. There was a small paddock between the farmhouse and some outbuildings, and it was here that the two rams got together. The first I knew of it was when I heard a terrific bang, just like a sledgehammer hitting a wooden post. I looked up from my sorting, and saw the two rams facing each other and stepping back until there was about three yards between them. Then, heads down, they each took two strides and bang, they were head to head. It was a terrible sound. I could not understand how they had both survived. I jumped up to grab one but the boss caught me and said "There's nothing you can do—you'll get yourself killed." By this time all the shepherds and dogs were in a big circle around the rams, but neither man nor dog would get too close. This fight went on for about four or five minutes. They were big old rams with huge horns; at last one got his horn entwined in the other's horn and threw him right over his back. I was shocked to see that in doing this he had broken his neck. Nobody, even then, would catch the victor, so we let about twenty ewes into the paddock and then ran them all into a small pen where we were able to pull him out and shut him in an outbuilding. From that day I have never trusted a ram; I think they can be worse than bulls.

IT WAS ALWAYS COSY in the farmhouse kitchen, with the settle on the back wall and a large scrub-topped table. Mr Middlemiss's chair was to the right of the fire, Mrs Middlemiss's chair to the left. Also to the left of the fire on the hearth there was, most of the time, a large bowl (similar to the one in the bedroom with the wash jug in). This was full of milk with rennet mixed into it to curdle the milk for the cheesemaking. After a while the milk would turn into curds and whey. The whey was then strained off to go for the dogs. Salt was added to the curds and mixed in. It was then put into a muslin bag and hung up to drain. These bags were hanging all around the kitchen in various stages of drying out. When the curds were hard and dry, the bags were put out on to the windowsills of every room of the house! They stayed there, being turned occasionally, until they were mature. A small hole was bored into the cheese; if the core was nice and crumbly, it was mature; if it was like dough, it was not yet ripe. The smell of this cheese was one of the distinctive aromas of the house.

A far better one was the smell of new baked bread. Mrs Middlemiss was a superb cook, and her bread and scones melted in your mouth. These smells pervading the house have stayed with me over the years, as has the smell of oil lamps. Trimming the wicks to get the best possible light was an art, but even so the light was very poor.

The two domestic jobs I had to do were to churn the butter and separate the cream from the milk. Using a separator was fine; I just had to start winding the handle very slowly, build up to a fast speed and hold it there until the milk had run through. The machine was a jumble of about twenty interlocking cogs; the milk ran through between them and was pressurised by the teeth, which for some reason forced the cream and milk out of different spouts on the side. But the butter was a different story. On some days it 'came' in no time at all, on others it took two or three times as long. It was strange how it took its time when I wanted to be away. I could not leave it; I just had to keep on going until I heard that thumping in the churn as the butter dropped from top to bottom, then take out the bung and let the buttermilk run out into a bucket. Then Mrs Middlemiss would take over to salt and pat the butter into shape.

Another job that I did not like doing—but enjoyed the result of it immensely—was, twice a year, killing the pig. A travelling slaughter man did the killing, but once the pig was killed and bled we had to set to and scrape all the bristles off with hand scrapers and boiling water. After this the pig would be gutted and hung in an outbuilding for three or four days to 'harden off'. All the insides were eaten. The boss said that the only thing that was not eaten was the squeal! The food was endless: liver, brawn and sausages (the sausage skins were made from the intestines!). One of the best bits was the chine, that is the backbone; it was a lovely sweet roast. A lot of these cuts are not available now as the pigs are killed so young. The pigs we ate would weigh twenty-six score (520 pounds). Today's bacon pigs weigh ten score (200 pounds)—much less mature. This makes a tremendous difference to the taste. Also our pigs were cured with salt; today's pigs are cured with brine so the bacon, instead of being nice and dry and fresh, is soggy and wet. When today's bacon is in the frying pan, it will spit a lot in cooking; this is water in the hot fat. And the white scum foaming on the rashers is brine. We had none of that.

There was always too much fresh meat for one family on pig-killing day so most of the homes in the village got a bit of our pig. This, of course, meant that we also received pork from the others when they killed. The sides of bacon and the hams were cured in the cellar with salt, which was rubbed in until it would not take any more. It was then put up in the 'cratch', a wooden frame that was suspended from the ceiling on large hooks, in the middle room to dry.

We enjoyed our food. You can't beat traditional home-bred country food, especially if you have reared it yourself.

We always had milk, cream and eggs to sell. The milk was from our Shorthorns. These were gentle beasts with colours varying from red to strawberry roan. They seemed to belong; they were contented on our herb and flower-rich limestone pastures. When in milk (i.e. after they calved), they would be milked in the mistle (cowshed) at home. Later, when they had finished milking and were in calf, they would be moved away to one of the field barns. Some of the farmers would milk their cows in the fields away from the village. They would set out with a 'back-can' carried on their backs in a harness like a rucksack. The can usually held about three gallons (you could buy them to hold two, three, four or five gallons), and some were shaped to fit one's back like a brandy flask. They were quite comfortable when strapped on. They also carried a milking pail over one arm, a feed bucket with a bit of corn

in it on the other arm and a milking stool with three legs in
one hand—and all while riding a bike! They would go about
a mile to the field, take off the back can and leave it at the
gate with the bike. Then they would strap the milking stool
on to their 'bum' so the three legs stuck out behind and
approach the cow, quietly put the bucket of feed in front of
her, sit down and milk her into the milking pail. Dead easy,
no trouble. I have even seen this done without the bucket of
corn. The farmer has quietly sat down and started to milk,
and the cow has just carried on grazing. if she moved forward
to graze, it was just a case of moving along; what made it so
easy was they didn't have to keep lifting and moving the
stool. The cow would often stand still and just chew the cud
whilst being milked. The milk was then poured into the
churn, and the next cow would be done. We used to love the
creamy milk produced from our rich pasture.

LIVING AND WORKING in Kettlewell in 1946 was very different
from today. It is still a lovely place, even with all the people
that now roam around the village and over (and under) the
fells, but when I first went there, there was no mains electricity, although many outlying places were being supplied by
'the grid'. I remember helping to protest at not having electricity by placing big stones on the valley side spelling out
"we want the grid" and painting them white. They really
stood out. One of Mr Middlemiss's sons, Norman, used to
drive the Dales bus up from Skipton, and he made sure everyone saw our handywork.

At this time all the lighting was by oil lamps. They used
blue paraffin, which was bought at the travelling shop,
Harkers of Grassington. This big van sold everything. It was
like going into an Aladdin's cave. If only we had had a bit
more money! The only things I used to buy were rabbit snares.

I caught a lot of rabbits—there were hundreds about. I
once set 100 snares in the lane going up to Hay Tongue and

on my way back down I had caught thirty-two rabbits! The only trouble was that there was only one buyer, a Mr Binns. He came in his old blue van, but he only paid tuppence (less than 1p) for each rabbit. That came to 120 rabbits for £1. Two of our dogs were good at catching rabbits. They used to go along a wall, one on each side, until they found a rabbit sitting in the middle of it. Many rabbits sat in these walls, sometimes two feet off the ground. It was then just a matter of reaching in and pulling it out, because the dogs would have caught it if it had tried to run away.

Kettlewell was a small village, with about 200 people, two shops (one with the post office), three pubs and a bank—the Yorkshire Penny Bank—and the youth hostel. The youth hostel brought a lot of outside people to the village. The Warden was Mr Gummerson. He was a character. He had taught his dog, a black Labrador, to 'talk'—well at least to count. Many hours were passed playing dominoes with it. It used to bark a number of times to tell Mr Gummerson which domino to play. They say he used to beat very good players! We would often find them in the King's Head, the pub run by Mr Robinson. There was always a big fire and a fine atmosphere. Mr Middlemiss used to go there most nights, and after closing he would go to the back room for supper with one or two of the locals.

A strange thing happened there one night. Mr Robinson had just locked the front door and we were going to have supper when there was a loud banging on the door. Standing there was a young man, very dirty and obviously exhausted. He struggled in and sat down by the fire, saying that he had just crawled under Great Whernside! We thought he must be mad. He said that he had gone down Dowker Cave, north-east of Great Whernside, and had come out in the middle of Kettlewell Beck, which is south of the moor. Well really! This was 1946 or maybe 1947, and it seemed impossible. Now I understand it is done all the time by groups of people. At any rate he got a free pint of beer and a whisky out of it!

Another strange thing happened at the King's Head. One day a man on a motorbike came in through the front door, and ran into Mr Robinson, who was just coming out from behind the bar, breaking his leg. It's obviously a dangerous game, being a landlord!

I once went upstairs at the King's Head and found all four walls of one room covered with paintings by L. S. Lowry. I thought they were wonderful, with their sad-looking matchstick men, thin figures with bent backs, the mills and back-to-back houses, and smoke everywhere.

I used to love going to the pub at night just to talk and play darts, but I have never had a pint of beer in my life. I always drank ginger beer.

Norman not only drove the Dales bus but was also the gravedigger. One day the vicar had gone down to Mile End Farm on a visit and as he got out of his car at the farm buildings (which are at the bottom of a hill) a cyclist came very fast down the hill, ran into him and killed him! He was to be buried on Saturday afternoon, and by Saturday morning Norman had not started digging. So Mr Middlemiss told me to go and help him; this was normal, as Norman usually left things to the last minute. In Kettlewell graveyard there is a layer of rock about four feet below the surface, and the procedure was that we dug down to this level and then someone from the quarry would come and blast the last two feet. Buried in the next grave but one was Cutlife Hines, a famous author whose family were our landlords. He had a wonderful piece of natural limestone for his headstone, and many people used to come to look at it. On this Saturday morning there was an old couple looking at this grave. As we had got down to the layer of rock four feet down. Norman said to me, "Right lad, brush it out and that will do".

At this the old lady turned to the old man and said, "They don't put them very deep here, do they?" Norman, who was sitting on the edge of the grave, looked up at her and said, "It's OK, lass, I haven't had one get out yet."

An event to look forward to in winter was the village hop or dance. After this we would come home late and stir the fire in the kitchen into life. As this was a Yorkshire range it had bars in front of the fire with the oven on the right and a hot water boiler on the left. We would cut a slice each from the home cured ham on the 'cratch' and put a big plate on the hearth; then with a toasting fork we would hold the ham close to the fire bars and cook it with the fat dripping on to the plate. Two thick slices of home-made bread dipped into the fat, with the ham in between: what a sandwich! My mouth is watering at the thought even now.

Kettlewell had a cricket team, although they did not play during the war years. I can remember a team playing there from Ilkley, lower down the dale. I could only have been eight or nine years old at the time. The cricket field was right on the side of the Wharfe, just before going over the bridge into the village—a beautiful setting. The charabanc was parked at the side of the bridge and the players walked into the field to the small pavilion. I remember thinking it was nice to see these men in white playing, but the thing that really sticks in my mind was the roller being pulled up and down the wicket. There was a short-cut from the village to the cricket field, via stepping-stones across the Wharfe. It was quite wide but fairly shallow, and we had lots of fun going over these, especially if someone was coming the other way. I suppose that if the cricket had started again after the war those stones would still be there.

May Day was always fun. We would get the maypole out of Mr Wiseman's barn and put it up on the green, and the children would come from school and do the maypole dance around it with long ribbons which finished up in a complete pattern. It would then be reversed and totally unwound. It was good fun and it often went wrong, but at the end of the day, after many buns and much pop, the pole would be left with a multi-coloured, tightly patterned ribbon sleeve on it.

Another busy day in the village was the trials day—not sheep dog trials, but motor bikes and motor cars. About $1\frac{1}{2}$ miles towards Coverdale (a very beautiful place) was Park Rash, a very steep 1:3 gradient which was loose stone with quite a few hairpin bends in it. This was where the trials took place. I was a marshal (very important!). Also present were the St John Ambulance people (even more important!) and a large crowd of people. It was just a race or races to get to the top in the quickest time. They went up one at a time and were timed over a set distance. It was good fun, with tremendous noise, lots of dust, smoke, loose stones flying about, men on bikes rolling downhill and men rolling downhill while their bikes were still going uphill—in short, pandemonium! One kind soul let me have a go on his motor bike at the end. I didn't make the first bend. I used too much power, the front wheel came off the ground, the back wheel raced forward and pushed it higher and I went over backwards. The bike and I parted company, but we were reunited when it fell on top of me. I was lame for a week.

Then the cars would have a go. I did (and still do) like these pre-war cars. They had their own individual style and character. The men too, with their goggles, flying helmets and large gauntlet gloves and their big thick overcoats, were characters. It was always a good day and we always looked forward to the sight of all the old cars and motorbikes outside the Racehorses and Blue Bell afterwards. The trade brought by this event was good for the village.

The main place for relaxing in the village was the hall. There was always a game of dominoes or snooker going on, and of course the regular whist drives and dances. I was never good at any of these and although I never missed a village hop, I didn't spend much time in the hall. I always preferred to have a game of darts in the King's Head, or just sit in front of the fire and enjoy the company. I was always made welcome even though I did not drink.

Other days out included going 'beating' for the gents shooting grouse. I didn't like this much but the social side afterwards was always enjoyable, though I liked to get back to Kettlewell as soon as possible.

EVERY YEAR WE VISITED the Kilnsey Show. Today it is a very big event, but it was much smaller in my day. Early in the morning I would have to walk down there with a small flock of our best sheep off the moor, usually about fifteen or so. They were then penned up for the showing. We spent quite a bit of time sorting them out from the flock, but no time sprucing them up; they were just good examples of our Dalesbred flock. They did look nice with their coal-black faces and their two white stripes, one each side of the nose and just a white surround to the nose and mouth. Their legs were black but sometimes they had a few white spots on them. I was very proud to walk them down the road to the show.

The highlight of the show for me, however, was the stone walling competition. Although the men were all using the same stone it was surprising how many different styles were produced; indeed it was said that going round the Dales you could often name the builder by the style of the wall. I would watch them all throughout the day, and, fired by what I saw, would go back home and try a different technique; I was still no good at it, however.

At the show there was also a sheep dog trial. Strangely enough I never could get interested in this. I think that was because it took place in such a small field and I was used to working on the moor where there were no confining walls. This was to change many years later when I became a keen competitor. Another competition involved throwing a stone at a small tree growing out of Kilnsey Crag. It was right up on the overhang, and the competitors would stand over the road by the beck and take turns to throw. It sounds a daft thing to do and not very good for the tree, but I only saw it hit once in three years. Most people could not even reach the crag, never mind the little tree! Once I saw a man nearly hit it while standing in a barrel! He was a powerful man, believe me. There were a lot of little competitions like this for bets, such as seeing how many 56 pound weights a man could pick up on a shovel, or trying to carry a wool sack on your back. This weighed only about 225 pounds, but it was the size that beat you. It was about eight feet long, four feet wide, and round like a cylinder; it was like trying to control a drunken man.

Once, after the Kilnsey Show, I was riding home to Kettlewell when I found a fox in front of me. He was about five yards away, and every time I put on a bit of speed to catch him, he just trotted a bit faster. When I got tired and slowed down, he also slowed down. We went on like this for two miles, then less than a mile from Kettlewell he turned off the road up into the crags on Middlesmoor. He never ran and took no notice of my shouting whatsoever. Come to think of it, I have never seen a fox run very fast, even when closely followed by a pack of hounds. They just amble along without getting out of a trot, then lengthen their stride.

WE USUALLY HAD about five dogs, all border collies. They came from people around the area and the boss would often bring one home from Skipton sheep and cattle market. It would immediately be taken into the small fold by the cowshed and put through its paces. Then either it would either go into the dog pens or I would never see it again.

Mr Middlemiss was a good judge of a young dog, as I was to learn. One day he arrived with a tri-coloured bitch called Brandy—black, white and tan. In the fold she ran straight at the sheep, scattering them in all directions and grabbing one by the fleece, dragging it for many a yard before letting go. We could not catch her. I thought that would be the last I would see of her, but the Boss said, "Put her in the dog pen." She turned out to be a cracker, the best dog for miles around.

The dog training at Sunter's Garth was not the best. Mr Middlemiss had a very short temper, which had its disadvantages. First, he would chase the dog (although he never caught one); then he threw his crook at it (although he never hit one); the crook would break ("Hey lad, put another shank on my crook"), and finally I would have to catch the dog and put it away. That does not sound much, but it often meant following it two or three times around the village. I would meet one or two farmers and as soon as they saw me they would shout "I see you're dog training today". We still finished up with the best dogs around, though.

It is a sad thing, but we always had one or two dogs go blind. We used to think that it was just old age, but some of them were relatively young. We had never heard of progressive retinal atrophy (PRA). This is a disease that can be carried by a healthy dog. If he is used for stud, however, some of the pups will go blind whilst others will not, and in our day it was very difficult to find out which dogs had got it. All

Border collie owners owe a huge debt to the International Sheep Dog Society (ISDS), which took the decision to ban breeding with all dogs that were found to carry PRA when it was found out how to detect it. This sounds easy, but believe me it wasn't. A good many of the top dogs in Britain were carriers. Imagine not being allowed to breed from the best dogs, with all the training that has gone into them. Imagine all the income that was lost in mating fees. It was a big decision, and the ISDS was not at all popular. It is now, however, as a collie rarely goes blind today. There is now another disease called collie eye anomaly (CEA). This is not quite as bad as PRA, but it is being attacked in the same way by the ISDS, so it should soon be under control.

Another thing that has improved over the years is the feeding of dogs. There are many books written on feeding today, but in my early days the dogs got milk, skimmed milk or whey, flaked maize and pinhead oat meal. They really smelt, but they were quite fit.

As I HAVE SAID, the sheep we kept were Dalesbred, one of the many local breeds in my day. They were small sheep with black faces and a white line down each side of the nose. They had black legs. They were pure bred, and were 'hefted' on the moor. This means they were born and lived all their lives there—or most of the females did, as the male lambs were sold in Skipton market and went to lowland farms to be fattened for eating. We kept the best of the females, although a few would be sold because we would have had an ever-increasing flock if they were kept. We drew out the small, poorer ones, and they too would go to Skipton Market. Farmers on lower land bought them and they grew into good sheep on this more fertile ground.

We kept our ewes until they were five years old, and then they also went to Skipton market. By then their teeth were broken from eating heather, and if we had kept them they would have lost weight as they could not eat enough. These old ewes would do very well on good grass and would give their new owners two or three good crops of lambs.

The various terms for the different ages of sheep can be confusing. With us, a young female lamb was called a gimmer lamb. At about fourteen months she was shorn and became a gimmer. She would lamb at about two years old and become a first crop gimmer. At her second lambing she became a second crop ewe, then a third crop ewe, then a fourth crop ewe; then she would be sold as a draft ewe (i.e. drafted off the moor). Another system counted the number of times she was shorn: she was first a ewe lamb, then shearing,

second shearing, third shearing, fourth shearing and draft. Yet another system is based on her teeth. A ewe lamb will get its first main teeth at twelve to fifteen months. These are two broad teeth in the middle of the bottom jaw. Next come one tooth on each side of the first teeth, at eighteen to twenty-one months. Then one on each side again, making six at twenty-four to twenty-seven months, and the last two (one on each side again) at thirty-three to thirty-six months. So these sheep were called 'two-teeth ewes', 'four-teeth ewes', 'six-teeth ewes', and 'the full mouth'. As drafted ewes they were called 'broken-mouthed'.

There is a further complication: when a lamb is weaned from its mother it is called a hogg or hogget in Yorkshire, a tegg in the South. These are what I call young adolescents. They are all in flocks of one age or are the crop of lambs for the season. The females are called ewe hoggs or ewe teggs, and the males are called tup lambs—99% of these are castrated as soon as possible (today they are castrated with rubber rings within twenty-four hours of birth). Then they are called wedders, so become wedder hoggs or wedder teggs. The few that are not castrated become tup hoggs or tup teggs. When they have been shorn they become shearing rams, then rams. There are also local variations such as thieves (young hoggs) and chilvers (ewe lambs).

Coming into Kettlewell on the Knipe Road from Kilnsey there is a turn-off just before the bridge going up to Moor End (where Mr Middlemiss used to farm before coming into the village to Sunter's Garth). A couple of hundred yards up this track is the communal sheep dip, built by the County Council for the use of all the farmers in the village. We had to dip the sheep twice a year, mainly to control sheep scab, a very small mite that gets into the skin. It is very itchy, and the sheep would rub against anything solid, rubbing all their fleece off in the process. It is very contagious, and without regular dipping it got to be a severe problem. Dipping also controlled ked and lice. Before the widespread use of DDT in sheep dips, we used sodium arsenate in carbolic. We used to dip twice a year to keep the flock healthy: once just after weaning the lambs and once in autumn just before tupping in November. We used to put a few buckets of boiling water into the autumn dip; this, with the carbolic, used to produce a nice, clean smell. The sheep would be 'slid' into the dip tail first and on their backs and left in the 'bath' for at least a minute. We had about four or five in the bath at once.

The dogs sometimes also got keds on them. They are hor-
rible things, with two claws that lock into the skin. They will
not release—even if you pull them in half, the claws will still
be embedded in the dog. We found that the only way to deal
with them was to put a knob of lard just where the claws
entered the skin. They would then start to eat this and would
not stop. They would blow up to twice their size, like a
balloon, until they just burst! If we tried to pull them off, the
dog would have a nasty sore for weeks, but with the lard they
showed no ill effects. Some shepherds used to throw their
dogs into the dip, but I do not like that practice.

One day when we were doing a summer dip we had a
young lad helping to catch the sheep and putting them in the
dip. It was a lovely warm day and he was wearing a light
straw hat. As he was putting a ewe into the dip, his hat fell
off and into the mixture. It was still floating but full of dip;
the boss, quick as lightning, pulled it out and said, "Quick,
put it on before it shrinks". He did, but what a mess—he had
to go and have a bath.

The sheep had to be totally immersed in the dip. We used
a wooden dipping paddle, a long rake-like handle with a
wooden piece cut like the figure '3' fastened on. The middle
of the '3' was attached to the handle with the two ends
facing upwards. We would push the sheep under with the
middle; then we could lift it up with either side and guide it
out of the dip and into the draining pen. They stood in this
pen until most of the dip had run off them and back into the
bath. Dipping was one of those days when a lot of help was
always welcome.

Once, while dipping, I was using the paddle to push the
sheep under, when she got stuck across the bottom of the
bath. At the surface the bath is about three feet across, and
at the bottom about eighteen inches. The sheep had got cross-
ways on and upside down; it was stuck fast and I could not
move it with the paddle, so I jumped into the dip, which was

about five feet deep, grabbed the sheep and heaved it out. When sheep are wet through they weigh about 160 pounds. I had to push it up the ramp and into the draining pen, and I was soaked to the skin.

Some farmers used to dip the lambs that were going to market with 'bloom dips', which gave them a coloured fleece. They would stand out from the others and look good. Then of course everyone started to do it, and it really got out of hand, with red and yellow sheep all over the place. I am glad this practice has stopped; we now have no distractions and see the sheep for what they are. In fact it was the Wool Board that stopped it by condemning the wool as unsaleable to the mills, so anyone sending in wool with bloom on it did not get paid.

Another practice that was stopped by the Wool Board was washing sheep. This was always fun. Before shearing, a dam would be made in the river and the sheep would be driven into it and made to swim across. In this way all the grit and dirt would be washed out of the fleece and it would be a lot better to shear. There used to be a premium on this wool but it was removed, so that washed and unwashed fetched the same price. We still used to wash, however, until a few years later the price for a washed fleece was set at less than that for an unwashed one, and the practice stopped at once. I think that too much lanolin was being washed out of the fleece with the dirt.

We always said that we could tell our own sheep anywhere, even if they were mixed up with others, and up to a point we could. As I have said, our flock were 'hefted', born and bred on the moor, where they stayed for the rest of their lives. Every now and again, however, we would have to bring in outside rams to introduce fresh blood so as not to get too much inbreeding. It was these rams that used to stamp their mark on our sheep. It was not only the slight difference in the colourings and markings—like white faces or more or less

white on their legs—but also their size and shape. For
example if the boss thought our sheep were getting a bit small
he would buy a good ram from a farm with better land
(as the sheep would grow bigger on more fertile soil). Very
occasionally he would even buy a different breed, like a
Swaledale, and use him for two or three years before going
back to Dalesbred. One year he bought a Derbyshire
Gritstone ram. I thought he was a wonderful sheep, so big
and a fine upstanding ram. He improved our lambs for the
next three years and we usually got top prices for them at
Skipton Market.

This also meant that our breeding lambs, which were
kept on the moor, had slightly different markings from those
of other farmers who had stayed with the pure Dalesbred,
and that is one way we knew our own sheep. It follows, of
course, that if one of the eight farmers in Kettlewell had used,
say, a Swaledale, one had used a pure Dalesbred, we had used
a Derbyshire Gritstone, and the other five had used different
crosses over three or four years, each farmer's would be
slightly different, even though they were all basically
Dalesbred. Even so, every year we had some new crop lambs
that we could not identify and we had some fun when all
eight of us claimed these 'waifs'. Soon after the war an ex-
army man came to live in Hagg Dyke and at lambing time he
would go out on to the moor and mark all the lambs with his
mark! Where we used to average less than one lamb per ewe,
at sale time he managed about two per ewe. He soon left the
area. The eight farmers in the village would never have
dreamed of marking someone else's lambs as their own, even
if they had had a disastrous lambing time and could really do
with a few extra.

Keeping the sheep on Great Whernside under control
meant checking them whenever the weather would let us,
seeing that they had plenty to eat, and driving the local strays
back on to their own moor. The food supply (that is, the

heather) was controlled by the sheep gates, as I have said. If there were too few sheep on the moor the heather would grow old and tough, and the sheep would find it hard to break off and eat. It would also break their teeth, so shortening their useful life. If there were too many, the heather would be killed by not having enough leaf to recover from constantly being eaten. It was our job to keep this balance, so if the heather was getting too old and tough we would allow more than five ewes through the gates. If, on the other hand, large areas of heather were dying, only four ewes were let through each gate, the number being determined each year. It was an art to ensure that there were no major fluctuations, which would mean that the regular age groups were difficult to maintain.

We always had thirty or so strays that we could not identify, and when we had a gather, it was my job to take them down to a farm in Coniston. The shepherds from all the surrounding area would do the same, all on the same day. You can imagine what a day it was! It would start early. The strays would have been left down by Scargill House on the way to Coniston. I would be dressed as if I were going to the moor but instead I would head down the road to Coniston with my little flock in front and dogs racing backwards and forwards, as I came across open gates and turnings, both down to the river and up to the moor. Thank goodness there were not many cars about—it would be a nightmare today. I always enjoyed it, as it was a chance to see a different bit of our dale. Most of the shepherds walked with their sheep, but some came by lorry from Nidderdale and other neighbouring dales. Everybody was there by about 10 o'clock and sorting would commence. It was time for good-natured banter. A shepherd examining a nice ewe would be told "It can't be yours, John, thars never had a sheep as good as that in your life." John, undeterred, would say, "Yes it's mine all right, look at the horn burn;

that's definitely mine." "Well," the first man would say, "tha must a lost it a couple of years ago then, for it to ha' got as big as that." On the other hand nobody wanted to claim a poor, weak old ewe. One shepherd was heard to say, "Nay, I'll non tak that home, I'll mek mesen ill diggin an oil (hole) for it."

This went on until all the sheep had been claimed; it wasn't often that an owner was not known. Not only did I have to know my own sheep, but also the sheep of the other eight farmers in the village.And most of the shepherds there were the same; it was quite remarkable how we managed. When we had finished and all the small flocks were tucked away in various buildings or small paddocks we went into the farmhouse for tea. This consisted of home-made bread and home-made butter, home-made cheese, pickled onions and a pint pot of tea. We really did eat well. I only met these shepherds on a few occasions each year, but we seemed to be so much alike and enjoyed each other's company so much that these meetings were very special.

IT'S STRANGE, BUT in my first two years as a shepherd I never did any shearing! At shearing time we would start off gathering the moor and bringing the sheep down to the 'in-bye' land, that is, the enclosed land. The way down and through the village was quite often fraught with danger because of the open gates and road junctions, where the flock could escape. But it was surprising how soon the dogs learned where to be and how to get there. One dog, Glen, was a master at this. As soon as I set off from Hagg Dyke she would disappear, and sure as houses there she would be in the first open gateway, even if it was halfway back home. Then when the sheep had passed by she would go to the next gate or road junction without any command. Even if there were two gates opposite one another and the sheep got away, the dogs were quick as lightning in getting them back. If there was a choice of two ways to go the sheep would always take the wrong one.

Shearing was usually done about June, by a gang who came from Ireland. We had quite a time with those lads, but my word they could work hard. Some of them clipped on the floor. Our sheep were quite small and some of these men were big and bent double. Others clipped on a bench. It was like a school bench, with the shearer astride one end and a wider section where the sheep was laid. Clipping was done with hand shears, and believe me, after a day's clipping my thumb was nearly dropping off! We used to like our sheep clipped with about one and a half or two inches of wool left. They looked as if they were covered in a sheet of corrugated iron. We reckoned that when they were out in the driving rain on the moor, this thick fleece, its ribbing would turn the water better and also be good insulation. Today the shearing machines cannot do this as they cut about half an inch from the skin, so I suppose the sheep are a bit colder for it. I am sure our method is still the best for mountain sheep.

In addition to the shearers, there were people who rolled the fleeces when they were cut. I am surprised how the wool holds together in the shape of a flat sheep, with the fibres at right angles to the floor. The fleece was laid down with the cut side on the floor, then each long side was turned into the middle. Then it was rolled up from the tail end. At the neck end a small amount was pulled out and twisted into a loose rope which was then wound around the fleece and tucked underneath itself. It was then put into a large sack which will held about forty of our small sheep fleeces. This was sewn up with string ready to go to the Wool Marketing Board for grading.

Like all our tasks, shearing was accompanied by 'drinkings'. Mrs Middlemiss would make sandwiches and scones with gallons of tea. The food was always so good we lived better than kings, in my opinion. We only had to see that huge basket and the tea tins coming through the gate to get a feeling of contentment. Mind you, the Irish shearers had an amber liquid of their own which gave them the same feeling I suppose. They even put it in their tea, but I could not spoil the good taste of my tea.

Shearing days were one of the highlights of the year. My day was spent bringing unshorn sheep into the catching pens and taking shorn ones back to the 'in-bye'. This was always a very noisy job as the ewes had been separated from their lambs for the shearing. This was done by running the ewes through a shedding gate. This consisted of a race, narrow enough to stop a sheep turning round, with a gate on a central post at the end which opened into one pen or another but not both. The lambs left in the field made a terrible row, which could be heard a mile away. They ran about in all directions like a headless chicken, bleating all the time. Quite a few got over the high walls, and some were on the tops of the walls. When I brought a mob of ewes back, the noise rose to a crescendo. It was turmoil for a while. It was surprising how quickly the ewe and her lamb got together again, even though the ewe was now entirely different in appearance. Once they were united, they went off as far away as possible from the others, trying to be on their own.

Shearing had to be done in fine weather. A shower of rain brought the whole job to a stop. This made it hard work, as the ewes and lambs had to be put together again, and then separated again when they were dry. Once a sheep is wet it takes hours for its fleece to dry, so it would be at least the following day before the job could be done again.

The shearing gang would vary in number but there would usually be about ten, with catchers who caught the sheep in the pens and passed them to the shearers. They used to go round all the farmers in the village, so they were about for quite a while, even if it did not rain. They would also come back to Kettlewell when they were shearing at Starbotton, a mile up the dale.

The nights during shearing were the highlight of the year. There were usually a couple of fiddlers and probably an accordion player with the gang, and they just loved singing and dancing. They would gather in the King's Head and in the village hall. It is hard to describe the pleasure and good, clean fun we had with those people. They enjoyed both work and the social gatherings to the full. It is a privilege to have shared those times with them. At some unearthly hour we would crawl into bed. The shearers would usually sleep in or on the wool sacks—if Mr Robinson at the King's Head could get them out of his premises. The following morning, as I passed the barn on my way to fetch the sheep for the start of another day, I well remember that mouth-watering aroma of bacon and eggs being fried. There they were with a pint pot of tea and still putting that amber liquid into it!

Since those days I have had great respect and affection for the Irish. I understand that these days the shearing gangs come from as far away as Australia and New Zealand. A New Zealander, Godfrey Bowen, came over in about 1948 or 1949 and amazed us by shearing a sheep in one minute. He caught his own and sheared 360 in one day.

LAMBING TIME in the Dales is May. Most of our ewes lambed on the moor, so we spent every day up there. They were long days and we covered many miles. It was a very precious time, and very important for the profitability of the farm. Nothing was spared in tending the sheep. We had to be able to walk through the flock with our dogs, and not upset the sheep. They would get to know us and we would know where most of the flock wanted to lamb. They would all have their favourite places and if they were upset by us or the dogs the chances of losing their lambs was very much increased. Several times during lambing a shepherd or his dog would surprise a ewe that was about to give birth, and she would jump and run away. If she had just lambed, she would leave the lamb and would not take to it again later, so its chances of living would be nil. On a few occasions I have seen a ewe run away with the lamb still half in and half out of the womb. It would finally be born with the ewe in full flight, but she never came back to look for it. So it was a case of letting the flock know we were there but not upsetting them in any way.

One of the best things that I ever had for this job was a spy-glass—a little telescope which cost me less than sixpence (2¹/₂p). It was a wonderful aid and it allowed me to do my job very much better. I would see a ewe on her own (at this time of year a sure sign of lambing) and watch her through my telescope. If she was not in any difficulty I would keep well away until she had not only lambed, but also licked her lamb. After this bond had been made with her lamb it was rare for a ewe to leave—

on the contrary, she would stay and defend it with her life.

Once a lamb had been born it would be up on its feet in about a minute, and ten minutes later I would have all on to catch the little beggar. These lambs could run nearly as fast as their mothers. We liked to catch them early to put our moor mark on them. It not only showed ownership but it also meant that if we saw it the following day we would know it had been seen to and was all right; we had enough to do without looking at the same lamb two or three times!

I would set off in the morning with my lambing bag, which contained three bottles of milk, a bottle of iodine, a bottle of glucose, a piece of 'lambing rope', a pen knife, a tin of 'raddle' (the paint we used to make our moor mark) and a tin of Vaseline. My precious spy-glass was always in the inside pocket of my jacket. I had another bag on my other shoulder, which contained my sandwiches and a bottle of cold tea. Every day for three or four weeks I would set off like this with my two dogs and my crook, and I have never been so happy in my life. I had to go, whether it was sunshine or raining—even once or twice in a snow shower—but it was just wonderful to set out. If the weather was bad I'd like to be off very early because a snow or sleet shower seemed to be just the time when the ewes would lamb! In fact we used to call them lambing showers.

The contents of my lambing bag were very useful. I could always tell if a lamb was not thriving; it would try to stand with all four feet on the same spot, so it had its back up. The first thing with a lamb like that was to fill its stomach up with the milk. Once this was done it would feel better and become a lot warmer. Once it was warmer, it would be livelier and start to chase its mother for her milk. My (cow's) milk would, I hoped, start off this chain reaction, thus giving new life to these lambs. An unhappy lamb like that would not roam far, so that the following day if it was still there, I would know that there was something else wrong with it, or more probably the ewe had no milk. This is where the dogs were so useful. I would have to catch the ewe or, since it could run far faster than me, the dogs had to catch her. Once they knew which ewe I wanted, it was so easy; they would just round her up. With one at the tail and one at the head she would have nowhere to go and I could easily \catch her. No fuss, no tearing about; just a quiet job well done.

Mostly the teats just wanted 'drawing' to get the milk flowing. They would become gummed up with the very rich, thick colostrum. But if she was dry, with no milk at all, I would mark her on the top of her head with the raddle so that later, after lambing time, she could be drawn out and sold. The lamb in the meantime would go into my coat pocket where it would usually stay, quite comfortable and warm, with its head stuck out. If I found more than two or three, I had to go to Hagg Dyke to drop them off in the hay until I had finished my day's work

We would try to get these lambs on to new mothers if we could, but we would always lose a few lambs. Sometimes it was just too cold when they were born and they would die before they could get enough milk. We also lost a few to foxes. I would take one of the dead lambs, skin it with my penknife and put the skin on to one of the lambs in my pocket. It would then be put back where the dead one had been, hoping that the ewe would come back, smell her own lamb's skin and allow the new lamb to take her milk. She often did, and would rear the orphan. It takes three days for the ewe's milk to go through a lamb. In this time the lamb will pick up the ewe's smell so that the skin from the dead lamb can be taken off; by this time the ewe will not know that it's not her own lamb. Sometimes I would skin as many as twenty lambs in a season, and most of the orphans would be accepted by the ewes.

All the new-born lambs had their navel cords painted with iodine, and glucose was given as a quick 'pick-me-up' for any lamb which looked a bit weak. It is quickly absorbed and gave it extra energy, helping it over a sickly period. Lambs soon pick up if you can fill their stomachs, and once they take a fill from the ewe they don't look back. I would get a lot of pleasure from seeing a sickly lamb improve over the days, especially one with an old bit of lambskin tied to it! I would recognise these lambs months later when they were on their own on the moor. Sometimes at Skipton Market I would say to a friend, "I remember that four-year-old ewe being born." I am sure he did not believe me, but I could.

The Vaseline was to coat my hand when lambing a ewe. My hands were quite big so I had to have some lubrication to get inside it, especially if the lamb had to be turned. Most births were straightforward and the ewe would lamb by herself. Some just had a foot back and when it was straightened the ewe would lamb easily, but they could not lamb if one or both of the feet were back. The lambs would die after a couple of hours and the ewe would be in danger of dying from blood poisoning if the dead lamb was not removed within a day or two. I used a fine cotton cord (now nylon) and slipped it on to the lamb's two front feet to ease it out so that it could not slip back.

The cold tea and the sandwiches were for me! I would eat when I felt I was on top of the job. It would be all go to see what had happened during the night, catch up with new births and any disasters, and get up to date with marking and dishing out the first bottle of milk or two. Then I would sit down with the dogs where I could see most of the moor and have lunch. Looking down on Angram Reservoirs or north to Buckden Pike was idyllic, with the different colours of the heather, the cotton grass and other grasses. There was still a nip in the air and quite a few patches of old hard snow in the hollows sometimes. The best view was over Coniston Moor

and down our own Wharfedale. I could see for miles; I never
tired of this panorama, so peaceful and so friendly. This time
spent just sitting was never wasted; I always saw something
that helped me with my work, like a couple of ewes going off
on their own to lamb—must mark the place—or there, not
half a mile away, those four ewes that always get away when
gathering. They have still got last year's wool on—two have
got their lambs, and will take some catching! Or a fox as
large as life; watch him, see where he goes to ground, then
later get the terrier men up to try and catch him. No, it's
never time wasted eating lunch.

WE HAD FIFTY or sixty inches of rain per year, which is quite
a lot when one is trying to make hay. When the lambing was
finished the 'in-bye' land was left with no stock on it, to go
up for hay. Even the cows had to go from the valley bottom
to higher ground, or on to land that was either too steep or
stony to cut. We had a small 'one horse' mower. It would cut
3' 6 " at a time, and was pulled by a wonderful Dales pony.
These great little horses are a cross between a shire carthorse
and a thoroughbred race horse. They have a heart as big as a
bucket, and are small with a bit of a feather behind each heel
(i.e. long hair from the back of the foot to about twelve
inches up the leg). Ours was called Champion, and was well
named. Mr Middlemiss used to do the mowing; my job was
to sharpen the knives. These were pieces of bar 3/8 " by 3/8 "
and 4 " long, with thin triangular sections about 3 " a side riveted to it. I had to sharpen two sides of the triangle on fourteen sections with a file while Mr Middlemiss went twice
around the field. He would then take out the used knife, put
my sharpened one in, twice around and change again. By the
end of haymaking I could sharpen a knife pretty well! The
ground was so stony that a section would often be badly
damaged or even broken off all together, so a new one would
have to be riveted on. When the field was 'down' (cut) it was

left to dry for two or three days before being turned. There were no machines (at least we didn't have one), so the turning was done by hand with a wooden rake. We would hit the thick butt of the swathe and lift at the same time; this would roll the swathe over for the underside to dry. There were about five of us so the job would be soon done, but our arms would ache a bit!

The next job was to 'dash' the hay about. That entailed walking between two rows or swathes, dragging and lifting the rake across first the left and then the right swathe, to spread and lighten the hay. It would now be spread like a carpet all over the field—the worst situation to be in if it should rain! From then it was a race to get the hay in. About an hour or two later the hay had to be raked into rows again. This time we would put about two swathes into one row—called 'a wind row'. If the barn was in the field we were working in we would sweep it up and pitch it into the barn. If we were in one of the many fields without a barn we would put the hay into haycocks. This was quite a skilled job because we had to ensure that they didn't fall over and that they could turn any rain. The bigger the haycock (some people called them pikes), the less hay was exposed to the elements. Making them entailed working around the field taking two or three 'wind rows' at a time (depending on how thick the crop was) and, with a hay fork, rolling up the hay like a snowball and building it into a haycock about 8 feet high. Then came the difficult bit: putting one big forkful on top to make a waterproof cap. When made properly these haycocks would stand for two or three months and turn the rain. They would go black, but an inch below the surface they would still be good green hay. Even so, we would want to get them into the barn as soon as possible. This was done by horse and sledge, which was just like a big toboggan pulled by chains. The haycocks were put on to it, taken to the barn and tipped off by the pitch hole. While another pike was being fetched that one would be stored.

The hay was meadow hay with lots of wild herbs and flowers in it. It was very scented and sweet, and very soft. Once again that big basket of sandwiches and tins of tea would arrive and we would have a well-earned rest.

In the Dales you will see barns in the fields. This enabled us to take the cows to the hay instead of carting the hay to the cows! It would take a long time to cart a field of hay to the village with only a horse-drawn sledge, whereas we could quickly sweep a field of hay into the barn when it was in the same field. The hay sweep had long tines that went under the hay and as the sweep was pulled forward by the horse, on two long chains, the hay would pile up against the curved handles. When the sweep was full we pulled it out of the swathes and took the load to the barn, where we just threw the handles up into the air. The sweep rolled over, head over heels, and the hay was left in a heap as the horse walked on. The barns were often built on a hillside, with the ground level at the back, much higher than the front. Inside the barns there would be tie-ups for six to eight cows and usually some calf pens on the ground floor with a hay loft over the top. There was a pitch hole in the hay loft, which was on the side which faced the hillside, so that when the hay was loaded through the hole we were often pitching either on the level or downhill. With the hay over the top and the cows tied up underneath it was lovely and warm in these barns even with a blizzard going on outside in midwinter. Even the old oil lamps gave a warm glow which seemed to be welcoming.

There was also a small bay for bedding. For this we used bracken cut by scythe in autumn from the fellside. We had a small patch that could be cut by horse mower, but generally the ground was so stony that damage would be done to the knives, and it was better to cut by hand. It's quite surprising where we used to get to with our horse and sledge. The bracken was cut about two miles from the barns. A big load was tied on to the sledge, and the first part of the journey would be over rough fellside. Then we went through the village and down to the barn on the road. The sledge runners made a lot of noise and left two white lines down the road. It was a dusty old job unloading at the barn.

The only other job that had to be done before the winter quarters were ready for the cows was to spread the 'midden', a huge heap of manure outside the barn from the previous year. This was one job that was done with a horse and cart rather than by sledge. The muck was loaded on to the cart, which was driven straight along a wall side, about six yards away from it. A small heap of manure would be dragged off every twelve yards. By the time all the muck had been cleared from the barn, the field would have all these heaps of muck in straight lines right across. Then we would set to and spread the heaps by hand fork until the field had a carpet of manure all over it. During the winter this would disappear into the soil and in the spring the field would be a nice emerald green, usually standing out from those that had not been mucked. We used to go around the fields in turn but we never had enough muck to do them all. The field with the barn in it got the most, as we wanted the thickest crop of grass as close as possible to the barn. We were always worried about rain, and the nearer we were to the barn the quicker we could get the hay in.

ONE DAY IN AUTUMN 1946 when up on the moor shepherding, we came across three dead sheep together. They had been worried by something—we thought a dog. It is quite distressing to see sheep that have been killed in this way. They had obviously been terrified. We could see from the way they were lying like rag dolls, bits of wool everywhere. It's strange—we rear sheep to kill to eat; we kill pigs, hens, rabbits, grouse, bullocks and many other things, but it is not like this. We kill foxes to protect our lambs, but it is not like this. It seems even worse than foxes killing lambs, because at least a fox is a wild animal. He has to kill to eat. I think this is the reason; we don't understand why a dog needs to kill.

The day did not go too well after our discovery, and we were pleased to get back home. We wondered whose dog had done the killing. We were upset that it would probably be a sheepdog, and that it might belong to one of our friends and neighbours. We went through all the dogs in the village and picked out possible culprits, so we were steeling ourselves to confront their owners. We knew that when the dog was found it would be shot. We knew that the farmer would be devastated, not only because of the loss of a dog but because he would feel that he had let his friends down. We knew that there would be no refusal, and no harsh words or falling out; it was just a nasty time for all of us.

We got down to the village in sour mood and called a meeting in the village hall. That was significant, because there would be no drinking or happy atmosphere. This was serious, and the King's Head was a happy place, not the place for our mood. When we were all present, Mr Middlemiss told of our find. Before we could get down to talking about individuals, someone said six dead sheep had been found on Coniston Moor that same day. Twenty had also been killed on Buckden Pike the day before; the man who told us thought that our meeting had been called to discuss this! At least one of our concerns had been removed—it was not just one of our local dogs. We thought now that there must be at least two dogs, possibly a pack! News travelled so quickly in those days; we had only been in the village hall about one and a half hours when the police arrived. They confirmed the other killings, but not the numbers. They had been in contact with the Army and were organizing a hunt for the following day.

At ten o'clock the following morning we were assembled on the bridge, by the King's Head. There were dozens of us: all the farmers from the village and Starbotton, a good many hikers from the West Riding towns and a truckload of soldiers. Similar groups gathered at Buckden, Coniston and Grassington. Each team had a walkie-talkie radio and Army operator with it, so that we would know what was going on with the other teams. We set off up the village, following Kettlewell Beck, then struck off up the green road on to Great Whernside pastures. That is a flattish moor that forms the boundary with Coniston moor. The first people stopped at the end of the green road and as we walked over towards Lodge Moor we kept dropping men off every two hundred yards or so. We got over as far as Stone Beck Valley, which runs down to Angram reservoir. We had a good line right across the moor, with men with shotguns every three or four hundred yards. Some of the soldiers had .303 rifles. We had wireless contact with the Buckden and Coniston Teams, and they had contact with Grassington. I was with a soldier near the end of the line at Stone Beck. He had a pair of army binoculars, about a foot long. He let me use them and I could not believe their power. The moor about a mile away looked so near I thought I could touch it, and the line of men that trailed right out across the moor looked about three hundred yards long! At that moment I set my heart on getting a pair as soon as I could; they would make my work so much easier. I finally bought a pair in 1975 for £500.

The line settled down quietly to wait. Most people sat down. The ones that were most alert were the gamekeepers, who never relaxed at all. After about an hour or so one or two were getting restless, and three hours later a few were making their way back home. Then we got a message via the walkie-talkie that a big black Alsatian had been seen on Buckden Pike and was heading over Great Whernside. This brought everyone on to the alert. It would happen now, just as the weather was closing in; we could no longer see as far as we could an hour earlier. An hour dragged by and then another—nothing. Then a stir came down the line; the radio man had just received a message that a big black dog had just gone through the line the other side of Coniston Moor! We hadn't seen it, and we were in a direct line in between Buckden Pike and the Coniston line. It must have gone through us; it was too much of a detour to get round us. We stayed in line until dusk then all trooped home, deflated.

There was no more news that night, but early next morning we were informed that the dog, a huge black Alsatian, had been shot in the station yard at Grassington that morning by a policeman. Later we saw a photograph of it on a long pole carried by two policemen on their shoulders. Its size had not been exaggerated. We were pleased to see the back of this beast. It had killed over forty sheep in the short time it had been up on the moors.

Sheep worrying is not a very big problem, but it is very distressing and causes a lot of heartache for both owners and farmers. We never heard who the owner of the dog was.

THE MOST IMPORTANT time of the year was the sheep sales—first the sale of lambs, then the draft ewes. This called for the biggest gathering of the year because not only were the ewes on the moor, but also their lambs. It was the most awkward 'gather' because the growing lambs had only known quiet dogs in the lambing period; now they were going to be driven into pens with a bit of force. The usual gang of eight would be out for the gather, with all their dogs. Some of them were new and on show for the first time, and would be watched with keen interest by the rest of us. If one did not work to command, a lot of advice would be given along the lines of, "That dog wants a bit of lead in one ear to straighten him up. Here you are [giving the unfortunate dog owner a twelve-bore cartridge]. Just pop that in your gun and put the barrel in his ear—that should do it," or "I think he only takes commands in Welsh." If the dog was really going well, it would be, "I'll give you five bob for that useless thing," or "That's a nice-looking dog you've got there, Jack, will it work sheep?" Once the gather was over and done with and all the sheep in (except perhaps those odd four that somehow or other always managed to dodge back), they would be separated into eight flocks again and go back to 'in-bye' land. They had to be taken home to be weaned because it would be impossible to drive the lambs away to Kettlewell and leave the ewes behind on the moor; not even our dogs could do that, so they all had to go home.

At birth the young female lambs would have been marked with a red dot on the rump. It was done very neatly with a small round piece of dowel, about the size of a pencil, dipped in raddle and carefully placed on the rump then screwed round and lifted off. When this drew out it left a neat round mark about the size of an old half crown. If I made a mess of this mark, the boss was not pleased. Indeed, I was proud of my marking; it spoilt the look of the sheep if the marking was sloppy and haphazard. These marks stayed on for a long time and you could not alter them once they were on. At a later date, when the tup lambs had been castrated, they were marked with a blue mark, put on in the same careful way, but in the middle of the back. Later still the female lambs that were going to be sold would have another mark, usually blue, put over the red one. So it was important to take great care with the marking. When they were in the sale ring we wanted them to look their best, and if they looked like a Picasso painting it did not help to sell them!

The first sale would usually be for wedder lambs, so the flock would be run down a race and shed with a gate, putting all the wedders into one pen and all the ewes and gimmer lambs into the other. This was quite easy because we just looked for the blue mid-back mark and turned them all the same way. Mr Middlemiss was a master at this and I can never remember him getting one in the wrong pen. This shedding gate saved us hours of hard work.

The next job was to take the ewes and gimmer lambs as far away as possible and put them into a stock-proof field that they could not escape from. We would then go back to the wedder lambs and hope that the stock lorry was not too long in coming to collect them. It was only sixteen miles to Skipton, which was good, as we didn't want them in the lorry too long. They would become too warm and soon start to get dirty, and we wanted them to look bright and fresh in the market. The boss and I would both go with them to Skipton, as we had a lot of work to do before they were sold. We usually had about 420 wedder lambs—two big lorry loads.

When they were unloaded at the market they went into about twenty-two pens. We now had to set to and sort them into three groups of evenly matched sheep, grading them as best, average and poor. We hoped that we would have about twelve pens of best, seven of average and three of poor. It was hard work, but when we had finished the lambs looked a picture, a nice even lot, and we were pleased to stand by them and answer questions that any prospective buyers wanted to ask. We always stood by our sheep; some farmers just sent theirs into the market and let the market men sort them out while they stayed around the ring until they were sold. I reckon that if you are afraid to stand by your own stock there is something wrong.

We would have to wait for the sound of the market bell, a hand bell that was rung for about two minutes before the start of the sale. When it rang a large crowd would assemble at the first pen. This was important as a guide to what the trade was going to be like on the day. The auctioneer would try his best to sell the stock well. Good lambs at this time were making about £5 each, so if, say, the bidding only got to £4 15s then we would be over £100 down on the previous year. That was a lot of money in those days. On the other hand, if they made £5 5s then the whole market would be buzzing and it would be a happy day. I might even get a bonus; the boss was pretty fair, and if he could give me extra money he would. So we were always on tenterhooks until we had sold our lambs. I used to be off home on the mid-day bus because there was not another until late afternoon. Norman the bus driver (who, as I have said, was one of Mr Middlemiss's sons) would always want to know whether his Dad would be in a good mood or not. Skipton was a busy place with lots going on, but I always liked to be going back to Sunter's Garth.

The sale of the gimmer lambs was usually held the next day. There were other sales later, but we liked to be in the first ones—the best buyers were at these. The next week they would be away to another sale further up country; so the best buyers would keep going north, sometimes right up to northern Scotland. There would always be local buyers, but we wanted to catch the national ones, who would buy twenty or thirty thousand sheep each. A lot would be bought on commission for the lowland farmers down in the South of England, and some would be bought for the big estates around the country. Yet others would be put into other markets. These men knew their trade well; they knew where our sheep would do well and where the Scottish sheep wanted to be. We were pleased to see them; they would remember our sheep from the previous year, and if they had done well they would buy again.

The following morning, therefore, the flock would be taken back down to the shedding race and run through again, this time taking out the ones with a blue dot on the rump. Now the ewes and the gimmer lambs that we were keeping would go back to the 'in-bye' field at the far end of the farm again, leaving about 200 gimmer lambs to go to market. After sorting them into even groups again I would have to catch the early bus back so that I would have time to take the ewes and lambs back up to Great Whernside, leaving Mr Middlemiss to sell on his own. He would know pretty well who would buy his sheep before I left in the morning; the same farmer bought them most years, or perhaps every other year. It was so much easier with good sheep to sell, as we could make friends with the buyers, find out how our stock had performed, and also know that it was being looked after.

We had one more sale day left, for the draft ewes. The 200 ewe lambs that we had turned on to the moor with some ewes meant that we could now sell 200 of the old broken-mouthed ewes and still retain the size of the overall flock. This is known as keeping the flock in regular ages; in other words we always had 200 ewe lambs, 200 shearings, 200 two-year-olds and so on, up to four-year-olds. They were the sheep that were hefted to our farm. They were our sheep, we looked after them, all the money from them was ours but we could not sell them—unless we left the farm, when we then sold them to the incoming tenant. They could not be sold to go off the farm—if they had been sold away to another area, the new tenant would not be able to keep sheep on the moor. If he tried to turn new sheep on to the moor most of them would just keep walking until they found better grazing, probably in Wensleydale or Nidderdale! But as things stood, he would have a ready-made flock for a start. We were, of course, allowed to sell all the ram lambs and the ewe lambs which were not required for replacements, also the old ewes that were surplus when the new ewe lambs came into the flock.

We collected the flock into a big stone-walled pen on the moor and looked at each ewe to check its mouth. All the bad-mouthed ones were put into a pen, and the others let back on to the moor. There would of course be 200 five-year-olds, but some of them might still have a full mouth, and some of the four-year-olds might have lost some of their teeth, so we would keep the five-year-olds and sell the four-year-olds. We would go home with a flock of 200 old ewes to sell, leaving the flock on the moor all tidy and in good form, ready for the rams to be turned into them in November.

THE RIVER WHARFE starts its life up on Birkwith Moor with a few little becks joining together and running down Langstrothdale, where it is joined at Beckermons by the Oughtershaw Beck. It's at this point that it really becomes the River Wharfe. From this point down to Hubberholm it is idyllic, with the old road running alongside, often with no wall in between, just an area of short grass. In summer, even in my day, many people came out to enjoy a picnic on the river banks. The river then changes a bit as it leaves the high fells and runs into some very good grazing meadows down below Buckden and Starbotton, then comes into our village past the school.

At a point above the school, up on the 'Cam', there is a wonderful echo. We used to shout across towards Middlemoor and the echo would come back loud and clear. I don't know what caused it as we were not in an enclosed space halfway up the Cam. After this, the river takes on its well-known form—wide, shallow and stony, with deepish pools and skirted by stunted, moss-covered trees. It is then joined by Kettlewell Beck, which has two sources, both on Great Whernside. The one on the north side is called Park Gill Beck, and runs down past Dowker Cave and Park Rash; then it is joined by a spur that rises on the south side of our moor, just before entering the village.

This junction and the first three hundred yards of the two spurs of the beck have provided me with a good many breakfasts. Believe me, there is nothing much better for breakfast than a freshly caught trout fried in a frying pan over an open fire. The trout that I caught were only small brown trout, weighing about a half pound or a pound. I didn't have a fishing rod so mine were tickled—caught by hand lying flat on a rock. I don't know why this practice is illegal because it's quite hard to do and there is no nasty hook in the fish's mouth. I once caught a rainbow trout in this way farther up by Park Rash; it weighed three and half pounds on Mrs Middlemiss's scales! There used to be a fish in a glass case in the King's Head, the largest trout caught in Kettlewell Beck, weighing just over two and a half pounds! It is a pity that I have not been able to claim the record until now!

One day I was tickling a trout and I just could not get it far enough out to grab its gills. I knew that if I tried too soon it would be gone like lightning. So it was a case of very quietly edging it farther out. I was lying full length on a slab of rock, my arm at full stretch down and my hand hooked under the fish. It was very uncomfortable and my arm was nearly dropping off with fatigue, but I stuck at it and finally secured my breakfast. I slowly raised myself up until my hands were on top of the rock, and then I saw a pair of boots right at my side not a foot away from me. It was the water bailiff. He held out his hand for my fish, which he put into his bag. He said I would be hearing from the police; then, no doubt, I would have to go to court. I didn't mind being told off by the local bobby, but going to court in Skipton really frightened me to death! In the event I heard no more about it.

Where the beck enters the village there was a dam which had sluice gates and was always full. The dam here at Town Head was built to supply water for a waterwheel which drove a turbine to provide electricity for the village. As the turbine could not supply enough electricity during the night, there was a building full of batteries, which were charged up during the day.

The beck then goes right through the centre of the village, past Sunter's Garth and down past the old waterwheel. Originally, this wheel drove the old Corn Mill. There had been a mill on this site since 1265; it finally closed as a corn mill in 1805. After 540 years of service as a corn mill, it started a new life as a cotton mill. Unfortunately it did not last long in its new role, being closed for good by 1822.

The beck passes the King's Head and goes on to the old stocks. It then passes the blacksmith's, and goes under a bridge into the Wharfe. The Wharfe passes the cricket field and then flows under the fine main road bridge. I remember one of the worst nights of my life on this bridge. It started one Christmas Eve when an American Super Fortress crashed on our moor. We had seen it fly down the dale with an engine on fire and were very surprised to see it coming back up again, still on fire. As it flew over Coniston Moor it was very low and it looked as if it was going to land. However, it flew on and straight into Great Whernside, only forty yards from the summit. We lads were kept well away, but Mr Middlemiss went up with Champion, our Dales pony, and the sledge, and brought eight bodies down with a big group of Americans. After the New Year the Americans came in force and brought down a lot of the remains of the aeroplane, Mr Middlemiss and Champion doing several journeys as well. The large black hole could be seen on the moor for years afterwards and bits of plane were strewn around like confetti.

During the Americans' visit they gave us long tubes of cigarettes called American Old Gold. Cigarettes were hard to come by in those years and, of course, cost money. I had never smoked, but these, being a free gift, must be all right. There were about forty packets of twenty in each tube, and I had two tubes. We went down to the bridge, and with great excitement started to smoke. I had just had my tea and some lovely fat bacon, and I have never felt so ill in all my life. Ian, my friend, said later that I turned green—and that was before I was halfway through my first cigarette. I never smoked again. It was a fortunate lesson for me to learn, and I am grateful to those Yanks for their present, as it stopped me from smoking before I had begun.

The Wharfe had a swimming pool just below the bridge, and in summer we spent a lot of time there mucking about, but I never learned how to swim. This was also the place where we washed the sheep before shearing. When it was washed, we had to wait for the wool to 'rise'. This was when it started to grow after the winter and the old fleece was lifted away from the skin by the new growth. This new layer of wool was easy to cut with the shears; the old wool had become matted during the winter so it was very tough to cut through.

The river then runs into the most beautiful stretch for about two and half miles. It does not matter at what time of the year; it is always serene. Some days when the spring sun is shinning and all the trees are breaking into full leaf, the water sparkles. The brightness contrasts with the shadows of the trees and the greens of the meadows, then later with the yellows and pinks of the flowers in the meadows. Even on dull days the sunshine and showers give it character, with the clouds racing past the sun creating ever-changing patterns on the fell. The two different places—the moor and the river—both respond to the elements in glorious ways. The trees and bushes growing along the riverside are not the tall, gracious ash and oaks of the fertile land lower down the dale, but they have a splendour just the same. They contrast well with the meadows and stone walls, with the purple moors as a backdrop.

These quiet stretches of the river were full of birds—kingfishers, linnet, yellow hammer, chaffinch, wren, the wagtails—with summer visitors like swallows, peewits or lapwings and, of course, the finest of them all, the curlew. The air was full of birdsong and they were well fed with the insects and midges along the river. There was also that other rare creature on the river, the fly fisherman. These men were often real country lovers. They knew a lot more about the Wharfe than I did, and were lovely people to talk to—as long as I didn't ask them how big the one that got away was!

The Wharfe that I knew well finished at another set of stepping stones by the Knipe Barn. There was a footpath on the eastern bank from just below Kettlewell which went to these stepping stones (now made of concrete). We had a field about a third of a mile past these on the west bank. It was a big field with a barn in it and we usually had some dry (i.e. non-milking) cows in it over winter, so it was a daily job to walk down there to feed and water them. I often came back over the stepping stones in the dark. It was at a time of year when the owls were hooting and there were plenty of bats flying about—quite spooky.

I WAS ALWAYS INTRIGUED by a large acreage of land at Park Rash by East Scale, known as the Deer Park. There was also another one further up Wharfedale at Buckden. It was difficult to imagine how they could keep deer on two such open areas of ground, but apparently the Percy family had

had a large tract of land around Langstrothdale that was used for deer hunting. Deer hunting was a great love of the Normans, and large blocks of land in the Dales were given over to it. If the land was hunted over by the King it was known as a royal forest, even though most of the land was not covered by trees. It covered the whole dale from the river to High Tops and it did not matter whether there were tenant farmers or shepherds on it, it was all hunted. The land hunted over by nobles and churchmen were called chases, so the Langstrothdale area would have been a chase. Buckden Deer Park would have been part of this chase.

The Deer Park was surrounded by a dyke, and on top of the soil thrown out from this dyke was a paling fence which was so constructed as to let deer come in from the outside but not allow them to escape once they were in. They were kept in this enclosure until the day of the hunt, when they were let out to give the hunters a good chase. This was also where the huntsmen's horses were kept. The poaching of these deer was looked upon as a very serious crime and if caught, the poacher would often be imprisoned or deported. It seems that it would be safe to assume that the Chase and Deer Park in Kettlewell would probably have been run on the same basis, but under the ownership of the Novices or even the Prior of Middleham Abbey.

One of the most striking features of the Dales is the stone walls. They were built from stone lying on the valley floor, so the size of a field was determined by the density of stone on the ground. If it was thin the field would be big, and vice versa. It is amazing how they keep standing after all these years; they don't often fall down—it is usually sheep or people climbing over them that brings them down. Mending the gaps is an art in itself. I was never any good at it although I spent many hours doing it. We called it 'glatting'.

Basically, the walls consisted of two walls three feet apart at the bottom tapering to fifteen inches at the top, with rubble in the middle and every so often a through stone. That was a big flat stone going right through and sticking out each side, which stabilized the wall. If a wall was broken, we would have to take it right down to the ground and start again.

I understand that the first walls that were built were very small enclosures close to the houses in the village. They were usually fat and bulbous, wide at the bottom and often incorporating large boulders that could not be moved. They were built so that vegetables would not be destroyed by sheep. They were never straight—usually curved or even s-shaped. The more orderly straight walls which formed larger squares or are found on long fields were probably built by monks! They built a lot of the long straight walls that run up the fell side and over the moors to form the boundary between lands owned by different monasteries.

KETTLEWELL WAS OWNED by Coverham Abbey at Middleham and by the Neviles, Earls of Westmoreland, in about equal parts. In the last few years of the 17th century it became controlled by Trust Lords, a group of people appointed by the freeholders of the village, because the Earl of Westmoreland lost his lands to the Crown when he led an uprising and was defeated;, The Crown had also gained the abbey lands at the Dissolution of the Monasteries in the 16th century.

Kettlewell's school has an interesting story. It seems strange to find a school so far out of the village, and it is obviously not an old building, having been built in 1883. The lead mining company which leased the rights to the mine at Kettlewell from the Trust Lords had a partner called Swale, a member of parliament who built a school for the village in the middle of the 17th century. He also gave some land and five sheep gates, the income from which would pay for the teaching of pupils and the upkeep of the school. As was the case with many schools in this era, it was overseen by the vicar, although it was not a Church of England school.

In about 1860 this caused a disagreement, as most of the mining community were non-conformists. The school was in a state of disrepair, so a committee was formed by the vicar to raise money to improve it, and he persuaded the Charity Commissioners to appoint him and his church wardens as trustees. He wanted to incorporate a Sunday School and impose Church of England rules. The old school was pulled down and a new one built on the same site. The vicar declared it a national school, and ruled that the schoolmaster had to be a member of the Church of England. Well, the villagers were up in arms at this and they withdrew their children. So there was a new school, but no pupils! When I became Moor Shepherd in 1945 this solid building was still empty and had been since 1876! In that year the Education Department ruled that the village must appoint a schoolmaster or a school board, but the vicar stood his ground and would not allow it. The village appointed a schoolmaster, who started teaching in a rented room. A school board was formed and they set to and built the school that now stands on the roadside by the Wharfe.

In 1777 the first Moor Shepherd was appointed by the Trust Lords, to look after the sheep on Great Whernside. I don't expect that his job was much different from mine. There was one big difference, however: at this time the lead mines on Great Whernside were getting into full swing, so there would have been a lot more people in the village and it would have been a thriving place. In those days there were the lead mines, a coal mine, the smelt mill, a corn mill, a smithy and also a market. Today the old mine workings remain and also the ruins of the smelt mill, but the corn mill (which later became a textile mill) has gone almost without trace and the blacksmiths is now a small craft shop.

Lead mining started in a small way about 1605 and ended 250 years later in 1886. There was a tale about two young lads who ventured into some old workings on Middlemoor. They found an entrance to a level that had collapsed, and after quite a time they got through to the tunnel. After travelling about 400 yards into the workings they came across a man sitting on some rubble with his back against the tunnel wall; he had a big beard, was dressed in old mine-worker's clothes and had a 'billycock' hat on. The lads were shaken by this, and could not get out quick enough; they ran to report this death to the local police. It turns out that the man must have been dead for about 100 years! He became known as Old Joe Billycock.

I never ventured far into these old workings myself, but I do find them very interesting. The ore was taken by packhorse from the mines to the smelt mill. It was carried in panniers, each horse carrying about 280 pounds. There would be a few teams working, so Kettlewell must have had over 100 horses working for the mines alone. The old smelt mill had a waterwheel, presumably to drive the bellows for the furnace. The small dams above the mill can still be seen, but the most remarkable thing still there is the draft tunnel going from the mill way up to the top of the Cam. It must be half a mile long and was built in 1868. It is about four feet six inches in diameter; the top radius is built of stone and is above ground level, but is covered with grass and looks like a green snake going up over the fell side. I have caught many rabbits in this tunnel, by walking up it and putting rabbit purse nets over the several outlets on the way up. The smelt mill was blown up during the war by the Americans trying out a 500-pound bomb.

There is a wide green road going from Kettlewell to the Great Whernside pastures, which finishes on top of a flattish pasture. This used to lead to a coal mine. It was not very deep and consisted of a 30-foot shaft, and the coal would be

dug out in all directions as far as it was safe to go. This kind
of mine was called a Bell Mine: the handle was the shaft,
and the bell the workings. When one area was worked out
they would just move along the seam and sink another shaft.
The circular spoil heaps are in a line following the seam
of coal. They are well grassed over now, as the spoil heaps
were of soft shale stone which allowed grasses to seed over
them. The best coal went for domestic use, and the next
best to the smelting mill. The poorer grade coal went to the
lime kilns. It was all transported by packhorse, as with the
lead ore.

The dale river soil was rich, with an abundance of flora
and fauna. As this was being enclosed, and farming became
organized, land became more valuable. So it came about that
the land in between the dale bottom and the moors became
enclosed with walls. The moors were very acid, being covered
with peat. The fell sides would not therefore support an
abundance of growth, without the acidity being neutralized
by the addition of lime. There was plenty of limestone, which
was converted into a nutrient that the plants could use
by burning it and putting it in small heaps all over the
new enclosures, as is done with farmyard manure. It was left
to 'slake', that is to weather for a while. Then it was spread
with a shovel so that the field looked as though it was
covered in snow.

Lime burning had been carried out for many years, as it
was also used for building before cement was used. The lime
kilns used for building were built on the site of the
monastery, castle or bridge that was being built. Then, when
the work was done, they would be pulled down. There are
other old lime kilns all over the Dales which were used
for agriculture. They were built of rough limestone and
positioned as close to the supply of limestone as possible, so
that they could be filled by wheelbarrow. They were square
or circular, about 15 feet across, and were about the same

height or higher. The 'core' was 8 to 10 feet across, and 6 feet from the top, tapering to the bottom to no more than 3 feet, where there was an iron grate opening to the arched tunnels from the outside. The cylinder and cone-shaped interior were lined with a covering of sandstone as an insulator to stop the fire breaking up the walls. It was the arched tunnels that gave the kilns their characteristic shape. It is also through these arches that the kilns were emptied.

The kiln was filled from the top. First a load of wood would be dropped in against the iron grate, then a load of limestone broken as small as possible, then a load of wood or coal, and so on until the kiln was full, making sure that it was always the coal or wood that was in contact with the sandstone inner wall. It was then sealed off and lit at the bottom. The lime would burn for about sixty hours, after which it would be raked out through the iron grill and carted away to the fields.

WINTER INVOLVES a shepherd in extremes. I was either out in several layers of clothes and frozen to death, or in the kitchen at Sunter's Garth snug and warm. Strangely, I found both very satisfying. Routine jobs took up most of the time: milking the three or four cows, mucking them out, feeding them, then giving them a sweet smelling new bed of bracken. It was good to see them in their warm 'mistle' (cowshed) chewing the cud and obviously contented. I feel sorry for the cowmen of today, with their milking parlours of cold concrete, no bedding, the doors left open to let batches of cows in and out all the time, and with acres of slurry and smelly silage. I did like the warm 'mistle' and the sweet-smelling hay, with bedding up to the cows' knees. The milking cows were in the village farmstead, the dry ones in two barns, one three-quarters of a mile away and the other over a mile across the river, so it took a while just to get round by walking—or on an old bike, if it wasn't snowing.

With these jobs done, there would be the never-ending work of mending stone walls, a bitter job on a cold day. We also kept an eye on the ewes; the rams would have been turned into the flock in early November. We would leave them in for six weeks, that is two heat cycles, so most of the ewes would go to the ram in the first three weeks. The rams would be taken out and put back into a well-fenced paddock so that we knew the date when the last lambs could be born. The ewes were now out on the moor but they would stay low down around the boundary walls, where there was a little shelter from the freezing winds and rain. The shepherding became a routine walk around this boundary, so we did not often get up to the 'Tops' in midwinter.

I had a few special days up there in the snow. If the snow had fallen in a still period there would be a white carpet all over, one or two feet deep; then sometimes there would be a hard frost, usually followed by a bright blue sky. This would be a day to carry on up to the summit after seeing the sheep. I used to go up just for the sheer pleasure of the walk, the view and the tranquillity of the high moor. Although he would never admit it, Mr Middlemiss used to enjoy this as well—otherwise, why did he come with me? There were no sheep up there; they were all down as far as they could possibly get. On more than one occasion I saw the shepherd from Little Whernside out on the moor on those days, so he must have had the same feelings, although he had the good sense to shepherd on horseback. We would perhaps be a mile and a half apart, but we never failed to acknowledge one another with a big wave. During this time the shearlings, that is the twelve to fourteen month-old sheep, would have gone away for the winter on to Mr Middlemiss's son's farm at Bishop Monkton, to let them 'grow on', and so they would not get in lamb. They would come back to the moor in April.

During the long nights of winter I spent a good many hours making a shepherd's crook or two. The head was made out of a ram's horn, and the shank from hazel. Our rams had huge horns, which were ideal for this job. The horn would have to be about four or five years old so that there was enough to bend into the familiar crook shape. This was done by heating it over the glass funnel of an oil lamp. When it became too hot to hold it would bend, so that it could be reshaped into the crook head. It is done a little at a time (about two inches) so it took quite a while. A ram's horn is far too thick, so it has to be filed and sandpapered down to a nice finish. Sometimes if there was enough horn I would carve a thistle on the nose. Then it would be glued on to a hazel shank which had been straightened over the oil lamp in the same way as the horn had been bent. Many hours work went into making crooks, and the horn would last for years. I have a crook that I made in 1954 that has been used for catching lambs every year since and is now just worn in nicely.

Today there are some wonderful crooks made with beautiful carvings on them. Some of them take 250 hours to do; they are works of art and can be seen on display at most agricultural shows.

I had a little job that started one winter: pumping the organ in the church. As there was no electricity in the village, the organ worked by bellows that had to be pumped by hand. The handle was behind the organ, where I sat out of sight of the congregation. I was never committed to any religion and my interest was purely financial—I used to get sixpence in the morning and sixpence also at night.

Village life in winter was lightened by the do's at the village hall, when the community spirit came to the fore. All the events were well supported, with plenty of tea, cakes and gossip.

Towards the end of winter we looked forward to the longer days and the sounds of spring, especially the haunting call of the curlew.

THE WINTER OF 1947 started out very well. The sheep had been dipped in October, and they were back on the moor and looking fine. As usual, the rams were put with the ewes in November, so for the next six weeks I had to go to the moor every day to move the sheep up to the summit. During the day they would slowly wend their way back down and sleep as low as they could get. Moving them in this way meant that the moor was evenly grazed and it would ensure that in spring the heather would send up new tender shoots evenly. If the flock were allowed to stay low down, the top of the moor would soon consist of old heather, which would be very tough for them to eat.

Moving the sheep also had another purpose: to mix the flock so that the rams would get into contact with as many ewes as possible. This was important, as each ram was expected to serve sixty to a hundred ewes in a six-week period. Ewes can only conceive when they are on heat, a period of two days. So it was important that they were in contact with the ram some time during those two days. If she missed, it would be another three weeks before she came back on heat. Putting the dogs around the sheep bunched them together, and whilst they were being driven up the fell side the rams would hopefully come into close contact with any ewe that was on heat. There were about eighteen rams on the moor with about 1,700 ewes on about 2,000 acres, so each ram had over 100 acres to himself. Without this daily gathering he could easily miss a few ewes on heat.

Before the rams were turned into the ewes, they were 'raddled': they were turned over and a dye mixed with oil was liberally spread between their front legs, so that when they served a ewe, some of the dye was left on the ewe's rump. We were then able to see how many ewes had 'gone to the tup'. The dye has to be kept fresh, and plenty was needed to leave a good mark, so the ram had to be caught every third day and the raddle renewed. So not only did we know how quickly the flock was getting in lamb, but we soon knew if a ram was not working, and he could be quickly replaced. Moreover, by changing the colour of the raddle every twelve days we knew the order in which the flock would lamb. We would start with a red raddle, then green, blue, yellow. By the end of the forty-eight days all the ewes should have gone to the ram. Any that were not coloured after this time would not be in lamb.

The next job was to take the rams out of the flock. We now knew when the ewes would start lambing, which would lamb first (red), which next, etc., and finally we would know when there would be no more lambs, as we knew the date when the rams were taken out. At least, that was the theory. Sometimes a ram escaped on to the moor a few days early, and served a few ewes before he was raddled. Sometimes one would work very well but would be infertile, so we would get ewes with two colours on the rump because they had come on heat again even though they had been served. Once in lamb, of course, they would not come on heat again. Just a few of these would probably mean that the ewe was infertile, which was not so bad, but a good many double-marked meant the problem was with the ram, and this meant a big loss of income at the year end. It was not possible to let the barren ewes have another chance to get in lamb, as they would not come on heat for another three weeks, and this would mean lambing would clash with the hay making, and also that the lambs would still be very small at sale time, so all the extra work would yield only a small return.

In 1946 the ewes had taken the rams well and the rams had been taken out in December. January 1947 came in cold, with hard frosts, then on the last day of January it started to snow. The ground was rock hard, so the snow immediately stuck and we got quite a depth the first night. By lunchtime it was over a foot deep. A week later it was six feet deep and we were cut off. No one could get in or out for the next six weeks. One of the most striking things was that all the walls had disappeared. I had never imagined the dale without walls. It looked so big and so wide. Another very difficult thing to cope with was the pure bright light. After a short time it was very hard to see; it was quite a strain on the eyes. The village shops did not stock sunglasses—they were not a good seller in our part of the world, especially in winter! When the sun came out, the whole dale looked beautiful, but sadly this beauty was to cost us dear.

We were, of course, used to snow and long winters, but right from the start of this one it was more difficult. The sudden quick build-up of snow to six or seven feet stunned us. The cows were in the cowshed, but we soon ran out of hay in the village, so we had to fetch it in from the field barns. But how? The hay was loose—we had not heard of hay balers in those days. It was hard work just walking, as the snow was up to my chest! We did have Champion, our

Dales pony, but we were unable to use the sledge, so we would fill sacks with hay and tie as many as we could on the pony, put one on my back, and trudge back to Sunter's Garth with at least enough hay for perhaps two days.

So the cows were snug, warm and well fed in their small cowshed. But it was very different for the sheep. They were, of course, out on the moor, up to three miles away. They had to cope with the snow most winters; they would scrape away with their front feet until they got through to the coarse grass or heather. Their needs at this time were not great; they would not be moving far, so they would conserve energy. They would huddle together for warmth under one of the moor walls, and would not come to much harm. Only in the last few weeks before lambing would they need more food, and by this time spring would be on us and there would be ample good grass around by then.

That winter, however, the snow came so thick and fast that they were completely buried whilst sheltering during the first day. They did not have chance to get out then, and by the second day they were incapable of getting out. By the end of the first week they were under at least six feet of snow, and in some cases they were eighteen feet under snowdrifts, a very serious situation indeed. Behind some walls there would be as many as thirty or forty ewes together. They would be able to move a little in their snow hole, but unfortunately the worst place for food was near to the wall, so they were trapped in the worst place possible for survival. The first two weeks we did not get up to the moor. We knew roughly where the sheep would be; there was nothing we could do anyway, and they would be all right in their snow holes. A couple of weeks after the snow had gone they would all be fine.

We were busy at home making tracks to the buildings and the barns in the fields. I remember one day I made a track to the farthest barn, only to find that the wind had blown the snow over, filling in my track behind me so I had to remake it all the way back, only to see the wind-blown snow close it again. The snow was just as level when I got back as it was before I had started. I was shattered. We were now climbing in and out of Sunter's Garth through the bedroom windows, about seven feet from the ground. We were lucky in Kettlewell to have a coalman in the village. Mr Wiseman had a large barn well stocked with coal, so we managed to get dry every night and we were really snug and warm in our thick stone farmhouse.

As time went on, we faced the problem of what to do with all our milk. Normally we would have just sufficient for all our demands, but we had five cows in calf. When these calved they were usually transported to Mr Middlemiss's son Mark's farm at Bishop Monkton for his dairy herd. Now it was impossible to move them even out of the cowshed, so our supply of milk kept growing. Mrs Middlemiss got to work making as much cheese and butter as possible. Cheeses appeared everywhere, in the bedrooms and the passages, both upstairs and downstairs. The front room was full of cheeses and butter. The dairy outside was full to the ceiling with both. Then she ran out of rennet! By the time the snow released its grip, we had some of the best calves that I have ever seen—they had had so much milk. The dogs were also in super condition. Very little milk had been thrown away.

By the fourth week, we were desperately worried about the sheep. We had started to try to get up to the moor, but it was horrendous. First there was the six feet of snow. It would have been very hard even if this were the only obstacle, but not knowing where the walls were, we kept walking into them. We just could not find the gateways! We also kept running into very deep drifts; it was impossible to get through them, and going round meant a much longer journey. It took us days to reach Hagg Dyke, and we had a few concerns about getting back home. We always took the dogs with us, even though it was difficult for them in such deep soft snow. I always felt that I was safe if my dogs were with me. It became easier to get around when after a few weeks it stopped snowing and the surface of the snow became hard and crisp with the frosts. We were able to walk on top of the snow, but it was very tiring when we occasionally went though this hard top layer, perhaps up to our thighs. At least the dogs were able to move at great speed over it.

When we finally got up to Hagg Dyke in time to do some work, we were not sure what to do. First we had to find the moor wall. We knew that most of the sheep would be right round it, and we located it when we saw a hole about a foot in diameter in the snow with steam coming out of it from the sheep underneath. Digging down and pulling the ewes out took quite some time, but at last we had about fifteen sheep out, with three dead ones left in the hole. The first sheep we dug out looked so pathetic; they were very weak and they just stood there, not wanting to move at all. We had no food and we were surrounded by miles of thick snow, a vast expanse of dazzling white. We felt helpless; we were willing to work all and every day to save the flock, but I felt so insignificant in this powerful, beautiful and overwhelming landscape. We just put our heads down and carried on looking for more of the flock. The dogs were keen to help us; they would start to dig in the snow and we would know that more sheep were there. By the time we had to leave we had found about ninety, not counting a good many dead. The following morning we struggled up with two sacks of hay, a pathetic amount, but the effort to get there was soon repaid by seeing the hungry ewes enjoying a feed for the first time in many days. It was disappointing, however, to see that some would not eat the hay; they just stood and looked at it. Many of them had not eaten hay before, and did not know what it was. It was such a sad time, and a hard lesson to learn: the way the lovely world of Great Whernside could turn into such a powerful cruel force.

We heard it about lunchtime: an engine, as clear as a bell. It turned out to be a small aeroplane glinting in the sun, the first thing we had seen from the outside world for a month. It circled round us; we gave it a wave and it went away. We carried on probing and digging, putting everything out of our minds except the job in hand. About two hours later it came back and we were surprised and overjoyed to see it drop two big trusses (square blocks weighing about 100 pounds) of hay down to us. We had a mad five minutes, waving and jumping up and down, with a warm feeling for the pilot and people who had thought of us during that terrible time that we were having. What a lift it gave us; work now seemed so much easier. The more sheep we could get out now, the more would be saved. We still had the problem of the ones that would not eat, however. Some were just too weak and we were too late in getting them out, but others would push the hay with their noses but not eat it. We found that once the ewes had chewed a bit they would soon be eating well. The ones that were not eating would watch for a long time before trying a bit. It was as if they had to be taught to eat, even though they were starving.

The days were now spent in a set routine. The boss would do the cattle whilst I went up on to the moor. The snow had settled down and was about four feet deep. The trouble was that the ground under the snow was frozen hard to about a foot deep so there was no warmth from below. This meant that it would take a long time to thaw. It was also getting more difficult to find the buried sheep, as a lot were under the very deep drifts. Every few days, whenever the weather was good, the little aeroplane would come over and drop us some more hay. It is one of the regrets of my life that I never met the pilot. We found out later that the National Farmers' Union in Skipton had set up this rescue mission. I have no

Despite all their help, however, we lost a large number of our flock. The thaw, when it came, was very slow. The tops of the wall soon came into view, but the snow lingered well into June 1947. Slowly, the groups of sheep that we never found came to light. It seemed as though they were every-where—we lost about 600 in all. Mr Middlemiss was left with half a flock of ewes that would be in poor condition at lambing time. Then, one day when we were up at Hagg Dyke, he looked at me and said, "I'm sorry, lad, but you'll have to go." Neither of us said much after that. I thought the world of Mr Middlemiss, and I am sure the feeling was mutual. Mrs Middlemiss was quite distressed, she even shed a tear or two. The call of the curlew was still over two months away. When it came I was not there to hear it. My new job was to take me away from Kettlewell, from the snug Sunter's Garth, from Mr and Mrs Middlemiss—but I can honestly say that my heart never left. If I am even now unhappy or sad, I think of those early days, and somehow I feel much better. The call of the curlew, my dogs and the sheep stay vivid.

One Sunday soon after my brief conversation with Mr Middlemiss, I was in church after pumping the organ when the vicar said, "I hear you are leaving us, Trevor. Do you know where you are going?" "No," I said. I was feeling quite useless, I felt as though I had let Mr and Mrs Middlemiss down. The vicar went on to say that before coming to Kettlewell he had been in Birmingham, and he had had a Sunday school teacher there with a brother who farmed in Herefordshire. He offered to write and ask her to find out if there were any jobs going down there. It was very kind of him to help, but I did not expect anything to come of it. It was a huge surprise when only three weeks later the vicar told me that he had had a letter to say that the farmer wanted a workman. I could live in, and could I get to Leominster station on April 23? That was two weeks' time. Heck, where a workman. I could live in, and could I get to Leominster station on April 23? That was two weeks' time. Heck, where the thump was Leominster? But it was fantastic. I had a new job somewhere 'down South'. And how big was the farm, how much moor was there, what breed of sheep did he have, were the fields surrounded by walls or hedges, how would I get there, how would I recognize my boss, should I take my bike?

Leominster

I DULY ARRIVED at Leominster station after changing at Crewe (in those days everyone changed at Crewe), and there waiting on the platform was Mr Downing. He greeted me with a big smile, and I immediately felt at home with him. He was wearing brown boots and leather leggings below a very smart pair of riding breeches, also a tweed jacket and a camelhair waistcoat. The only thing that I had in common with him was a flat cap. I put my case in his Ford motor car and we set off to Moss Hill Farm, Monkland, a journey of about five miles.

The first thing I noticed was that there was no snow anywhere, which was very strange. He said that there had been a big snowfall but that it had all gone by early March. I don't think he believed me when I told him that when I had left Kettlewell two days before, the snow was still four feet deep in many places. He told me that spring planting was over (I did not like to ask what they were planting), and lambing had started the first week of March and was all but finished. Indeed, on our journey we saw fields full of ewes and lambs in lovely green grass about four inches high. I noticed how enormous the sheep were; they looked so big and fat, with so many lambs, it looked unreal. Mr Downing said he was a little disappointed that his ewes had only averaged one and a half lambs this year—he put it down to the bad winter! I could not help but think of our poor ewes on the moor still a month away from the start of lambing and so thin.

There was no moor or heather at Moss Hill, just fields with big green hedges around them. How I longed to bring Mr Middlemiss's sheep down to this haven, just to give them time to get into good condition on this wonderful grass. The next shock was that Mr Downing had a flock of sixty Clun Forest ewes. Sixty! I had just lost my job because we were down to 600. It was a very strange world that I had come to.

In fact, this journey to Moss Hill Farm was to be the start of many exciting new experiences for me. Everything was so different. I did not see a stone wall anywhere. Many of the fields were ploughed. It was a mystery how these fields appeared, because looking out across the country it seemed to be all trees, just like a forest. As we travelled along, there were fields on each side of the road all the way, but looking back from a rise, it looked to be all trees again. Not only were there thick hedges around the fields but everywhere there were big trees, mostly oak and ash, but really very big. Another thing that I noticed on that journey were the black and white houses, each with a lovely garden around it. The distance between Sunter's Garth and Moss Hill Farm was about 200 miles, but the two were like different worlds. Both were farming communities, both were rural villages, but was difficult to find any two farming activities that were alike.

One thing that was the same, however, was the welcome cup of tea which awaited me at Moss Hill. Mrs Downing was also a very good cook, and I well remember the spread that awaited us in the kitchen. Also awaiting us in the kitchen was Bob, a sheepdog. I had hated leaving my dogs behind at Kettlewell, and already I missed them, so it was good to see a border collie there and I soon made a good friend of him.

The farm was 168 acres, just about half our 'in-bye' land at Sunter's Garth, with no moor. Every field could be ploughed, and was, except for some river ground which, although it was very fertile and flat, was likely to flood. Mr Downing grew oats, wheat, barley, sugar beet and mangolds. He also had the sixty sheep he had told me about, twenty pedigree Hereford cattle, a few pigs and hens, and six horses to do all the work. There was also another farm worker called Tom Lewis.

It was a time for fast learning. I did not know anything about growing corn or sugar beet. The world of pedigree Hereford cattle was a mystery. The countryside all around me seemed to burst with an almost tropical growth. Each field seemed to have a different crop, and they all had to be managed to keep on top of them. The work was very varied, and was done without tractors or pick-up trucks, just horses and manpower. It was also good fun, with time to go to the village hop and also the Three Counties Agricultural Show and the local ploughing match. And I must have been growing up, because I went out with a girl for the first time—at the age of eighteen!

My settling in period was smooth, but I was very homesick for my dogs and the high moor, and I had some difficulty with the different dialect. The very first morning I was asked to clean out the pigs, and being unable to find a wheelbarrow I asked Tom, "Where's wheel bara?" Tom asked me three times what I wanted. He finally went off and came back ten minutes later with the required wheelbarrow. Two months later Tom told me that he could not understand a word I had said, so he had gone to ask Mr Downing what I was supposed to be doing, they had worked out that I must need a wheelbarrow. I for my part had difficulty in understanding their sing-song accent. The words 'thee', 'thou' and 'yonder' were not used at all, and they did not understand my morning greeting of "Now then". It is a shortened version of "Now then, how are you today?" Sometimes I used to go as far as "Now then, how is ta", but it still brought blank looks.

I think the worst task that I had to do was to catch the horses in the morning. The idea was to go out with a halter over my shoulder and my cap full of oats. I would offer the cap of oats to the horse and it would come to eat. At this point I was supposed to take hold of the 'topping' (the top end of the mane) and slip the halter over the horse's nose and then over behind its ears, and gently walk it back to the stable. Although Mr Downing and Tom did this without a hitch every day, I could never manage to grab the topping before the horse had eaten all the oats so of course he would gallop away down the field with me in pursuit offering him the now empty cap. As we had six horses to catch, I was pretty well fed up before all the horses were stabled. This daily pantomime seemed to give Mr Downing and Tom a great deal of amusement. It was a contrast to Kettlewell, where all I had to do was open the field gate and Champion would follow you all the way home. One thing though was the same as at Sunter's Garth: I was still living in on the farm and I was still on ten shillings a week!

The farm was a part of the big Burton Court Estate, which belonged to our landlady, Mrs Clews. The First World War had claimed the life of Mrs Clews' only son, and Mr Clews had died before I went to Moss Hill. Many estates of this era had lost their sons and heirs in the war, and this was one of the reasons why the estate system started to break up. The estate had a cottage on the bridge in Eardisland, which was let to Mr Downing for a farm worker. There were five other farms on the estate, and in a corner of Moss Hill farm there was a cottage for the gamekeeper. Hunting and shooting were important, and around the farm there were small gates about four feet six inches wide, called hunting gates. They were for the riders following the hunt to go through if they were unable to jump the field fences or hedges. Some of the hedges had a strand of barbed wire to prevent cattle from destroying them, and it would have been dangerous for the horses to jump them.

The farm had a road running the full length down one side, and another going across it, cutting off the river ground. One boundary was the River Arrow, which runs through some of the most beautiful parts of Herefordshire. Whereas the River Wharfe is shallow and wide the Arrow is quite narrow and deep, with steep banks on each side. Both are lovely rivers and yet very different.

ON THE OPPOSITE BANK to our land was the Arrow Mill, which was run by Mr and Mrs Davis. It was a corn mill driven by a huge waterwheel. The farmers for miles around came to have their corn ground here. It was brought in by horse and cart: sometimes a dray, a four-wheeled cart pulled by two horses, one in the shafts and one in traces in the front. One used to come from Street Court with two horses pulling it and a third walking behind. This one was hitched up to help pull up the steep hill to their farm. It was quite a sight to see one horse in the shafts and two trace horses pulling as hard as they could up the steep hill. It also took a good waggoner to handle them and get them pulling as a team. I remember the first time I had a trace horse to manage—it was a disaster! The shaft horse started to pull before the trace horse, and would stop just as the trace horse started to pull. Then the shaft horse started to pull again just as the trace horse stopped!

There were always a lot of people at Arrow Mill. It was a busy workplace, but also a sort of community centre where all the local news was circulated. The mill itself was a work of art. The one waterwheel did many jobs: turning the huge grinding stones, working the hoist that lifted the corn from the wagons up to the top of the mill, working the sieves that took out any rubbish or stones from the corn. All the cog wheels were made of wood and had to stand a tremendous amount of pressure. The water ran under the wheel, a system called 'undershot', while a wheel that is driven by water going over the top is called 'overshot'. The big grinding stones were gritstone and came from the Derbyshire moors. They had to be 'dressed' quite often, with their grooves kept to a certain depth. They were made in such a pattern that the corn, as it was being ground, was forced to the outside. It was fed into the middle of the stones and travelled across them and fell into a collecting tray. The whole complex of the mill—the river, the mill race, the wheel, the shafting and belts, the creaking woodwork, the horses and carts, the babble of happy, often 'ribbing' voices of men at work—made for a lovely environment.

I had always been used to working sheepdogs. It is difficult to explain the pleasure I got from them. The thrill of whistling to a dog that is half a mile away on a wild moor and see him turn left or right on his two different whistles is simply wonderful. The notes of the whistle are traditional and have evolved over a long period. They imitate the calls of the curlew and lapwing, birds of the high moor. To tell my dogs to go left I gave the lapwing call, and to go right, the curlew. The order to stop was one long blast, and to make them walk on again I gave two short blasts. The only other whistle that was needed was one to make them look back or go back. Mine was one continuous blast, starting high, then going low and then returning to high again. If you go to a sheepdog trial you will hear many whistles for these five basic commands, most of which would not carry over a spacious moor but are quite adequate for a small trial field.

Another difference between the moor and the lowlands of Herefordshire is the sheep. The mountain sheep are small and light, quick on their feet, turning and moving with the dog some distance away. The lowland sheep are very much bigger and sluggish, hard to move—as we say, heavy. The dog has to work much closer to them to push them around. So although border collies are used for both, the strain within the breed has to be quite different. Most of the top dogs can work with both mountain and lowland sheep, but it requires a good handler as well as an exceptional dog.

This difference is my excuse for my first disastrous attempt at sheepdog trialling. I thought that as I was a shepherd, I knew how to handle a dog and sheep, so I did not need to practise. Old Bob was all right at home, perhaps a bit unreliable, but we always finished up with the sheep where we wanted them. When I heard that there was to be a sheep dog trial at Amstrey, which was only about five miles away, I asked Mr Downing if I could have a go with Bob. He was delighted and said he would come along and watch. We finished our work early on the Saturday morning, then set off in the Ford with Bob in the back. It was only at this point that I began to wonder exactly what I would have to do.

Although I had seen a sheepdog trial at the Klinsey Show, I had never really sat down to watch what was required. I must have been very naïve to think that I could do what was needed just because I was a shepherd. Still, it was too late to back out without losing face.

As we drew up in the field there was a tug-of-war going on, with a team on each bank of the river and the rope stretched taut over it. Someone was going to end up very wet! I had no time to watch this as right in front of us was a steep bank with a sheepdog trial course set out on it. We were not in a small field with a thick hedge around it, but on a bank, with the roadside hedge down the left hand side and no other hedge anywhere in sight. I did not have long to wait for my run. I had paid my entry fee of five shillings and was told what to do by the course director. I had to collect five sheep from about 400 yards in front of me, bring them down through two gates twelve feet apart, and up to me. Then I had to turn them around behind me and drive them diagonally away, out to my left and through two gates, then turn them to go straight across the field about 100 yards in front of me, through yet another set of gates. I then had to bring them back to my feet and put them into a ring of sawdust which was 30 yards in diameter, and cut one of the five out whilst the others were still in the ring. The dog had to hold it and I had to stop it from rejoining the other four, then collect them all together again and put them into a pen. The time allowed for all this was fifteen minutes. No trouble—just get on and do it.

I was called to the post, dog at the ready, and with a command of "Away" he was off like a good 'un, a good wide out run and behind the sheep—beautiful. He came on to the sheep a bit fast but just in the right position. My first touch of panic came at this moment, as I realized that the sheep were not our big heavy ewes, but very light Welsh mountain ewes. With Bob bearing down on them at great speed they took fright and tore off at an even greater pace. They were charging towards me like an express train, dead straight but fast, straight through the middle of the fetch gates. They roared on towards me, with Bob now left 100 yards behind. Straight past me they roared, past the judge's car, through the spectators, on down the field and though the gate on to the road and away into the far distance. The course director told me very sharply to fetch them back. I suppose I was lucky in one way because I escaped the embarrassment of having to face a critical crowd for long. Without a word, the boss drove me three miles along the road and we caught up with the wayward sheep. They were pretty well out of breath by then, thank goodness, but even so it was late afternoon before we arrived back at the field. I had hoped that the spectators would have forgotten that I had gone by then, but a big cheer went up as I slunk back into the field. There were big smiles all around, and I learned what a good bunch those 'dog men' were. One of them told me that up to the time I left the post I had not lost a point!

THE DAY-TO-DAY WORK of the farm was very varied. The sheep were kept to produce lambs, which were kept on the farm until they were fat enough to go for meat, whereas at Sunter's Garth the lambs had been sold off for other farmers to fatten. There, our flock also consisted of ewe lambs and one-, two-, three- and four-year-old ewes; here, Mr Downing would buy two-year-old ewes, lamb them for two years and then sell them, so every two years we had a completely new flock.

A new job for me was the topping and tailing of lambs— indeed I had never heard of the practice. 'Topping' meant castrating the lamb. It was done with a red-hot iron, which sounds horrendous, but in fact the lambs would be up and running around after just a few minutes. It was done to stop the male hormones from developing, as this gave the lamb carcass a strong, unpleasant taste. The lambs from Kettlewell were much smaller and reached the killing weight at a much earlier age, before the strong taste had time to develop, so they were not castrated. Tailing was also not done in Wharfedale. This involved cutting off the tail close to the body, leaving only about four inches. This was also done with a hot iron, and again the lambs recovered in minutes. This was necessary in all lowland areas because of the constant threat of maggots in a dirty tail. I cannot imagine anything worse than a young lamb being literally eaten by those horrid larvae of the blowfly. In the Dales we were too high for this to be a problem for us; I never saw a sheep with maggot until I came to the lowlands.

The shearing was done with a shearing machine. This had a hand piece like a barber's clippers, connected by a flexible tube to a gearbox that was turned by hand. This was another art that had to be learned. Being a shepherd I was expected to be good at it, but of course I had never seen one before. Hand shears, like the ones used at Sunter's Garth, were just used to 'dag' the ewes, that is to cut off any dirty wool around the tail. When I sheared a sheep with a pair of hand shears Mr Downing and Tom couldn't stop laughing, they said I had left more wool on the sheep than I had taken off! I, for my part, did not like to see all the wool taken off right down to the skin. I thought the sheep would freeze to death, but the climate in Herefordshire was much kinder to the stock.

I was also expected to know how much a lamb weighed just by looking at it. At four months these lambs were bigger than our ewes in Yorkshire, but I soon got into the way of estimating their weight and later took quite a pride in guessing it and then weighing the lamb, finding that I was less than a pound wrong in a lamb of about ninety pounds.

I loved going to the sheep fair in Leominster. This was for ewe sales, and it was not held in the market, but in wooden

pens set up in a field just outside the town (the fat lambs were
sold in Leominster market at the weekly market held every
Friday). Hundreds of sheep would be sold in the day, and
then the pens were all taken down again until the next year.
About ninety per cent of the farmers within a twelve-mile
radius came. There was a tea tent and a bar tent, and a
cattle lorry was used as an office for the auctioneers. I took
a lot of ribbing because of my accent. It seemed to give everyone much pleasure, but it was all in good fun and I made a
lot of good friends on those days. We used to go to every fair,
even if we had bought the year before and so were neither
selling or buying.

One man I got to know well, who was at the fair every
year, was the guinea man. He was a real character, striding
around with a dead straight back. He wore a top hat and tails
and always had a rose or carnation in his buttonhole. His top
hat was a bit the worse for wear, and his coat was literally in
shreds, but he nevertheless wore them with distinction. His
vocation was horses, and he knew all the farmers for many
miles around. If a farmer wanted a horse he knew where to
buy it. He knew who had a carthorse, a hunter, a trotter, a
racehorse, a trap pony, a child's pony or any other kind of
horse for sale. He would tell the farmer where to get fixed up
and if a sale was made the buyer and seller would both pay
him a guinea. Sometimes the deal would be made many
months later, and the vendor might be in mid-Wales and the
buyer in south Herefordshire. But if a sale was made, the
guinea man was always paid. On sale day he was always in
fine form, because it would be a good payday for him.

Many days were spent on mundane jobs like hoeing
mangolds and sugar beet. One day we were hoeing the sugar
beet, and a gypsy called Ivor was helping us. He had his van
down on the river ground with his lurcher dog (he called it
his dinner catcher) and a coop with about a dozen hens slung
under the back axle of his van. It was a lovely place to have

his van, with running water (which might also supply fish for his tea) and flat land which was good for his lurcher to catch rabbits on. His van was the old type with a canvas hoop top, but it was spotless inside. As Ivor was hoeing on this particular day, another gypsy came down the road with a brand new van, which had a fancy wooden top. It looked beautiful, newly painted in green with red and gold lining. Ivor had a word with him over the hedge, and found he wanted to sell it. He wanted £350 for it, and to our amazement Ivor bought it. He sold his own van to the man for £100. He picked up his old coat from under the hedge, drew out a roll of notes and paid the man £250 there and then. Tom and I could not believe what was happening. Then all three of us carried on hoeing as the man went off to the river ground, and we could see him and Ivor's wife emptying Ivor's old caravan. When it was empty, the man hitched it up and drove away.

I think the most boring work I had to do was 'thistle spudding'. We did not have chemical sprays to kill thistles in those days; instead we had a thistle spud, a tool with a long handle and a thin, flat, oblong piece of steel about two inches by three on the end of it, with a 'V' shape cut out of the top near the shaft. The front edge was sharp, like a Dutch hoe, and it was used to cut the thistles in growing corn. The 'V' on the top was for cutting larger thistles, called 'bull thistles'. We placed the thick stem in the 'V' and pulled sharply towards us. For the first few days blisters on my hands were a problem, but I soon hardened up with well calloused hands.

Another of my daily tasks during the winter months was 'suppering up'. We had six horses at Moss Hill, and every night at about 9.30 I had to walk them out of the stable to the water trough for a drink, take them back, give them a forkful of hay, sweep up behind them and give them fresh bedding for the night. I had started to go down to the village hall once a month to the dance because the daughter of the next-door farmer used to go, and I had seen her and spoken to her as she went around the sheep. I enjoyed these nights out, and it was a bind having to cycle the three miles from the village to supper up at 9.30. It was made worse because the stable was only a few yards from the house and the horses made quite a noise with their big feet and steel horseshoes on. I would try to walk them round the edge of the muck heap, but they still made a noise. This alerted Mr Downing, who would open the window and say, "Is that you Trevor? It's too late to be going back down there now, you had better get to bed." My heart would sink—I really wanted to get back to the dance to walk back home with my new girlfriend.

The other job that was a trial was filling the water tank at the top of the house every Saturday morning. This held 500 gallons of water and was connected to the pump outside the back door. I had to work the big old handle on the pump until the tank overflow ran; sometimes this would take over an hour, depending on how much it had to be topped up.

The river ground on our side of the Arrow was the part of the farm I loved the best. This was strange because my favourite place at Kettlewell was the high moor—what a contrast. The grass down by the river was so lush it was amazing; it just grew and grew, with thick wide leaves, emerald in colour. It was grass for cows and calves or big bullocks, and did they look good? The red and white Hereford cows with identical calves, the green grass and the different shades of the hedges and large trees, with big white sheep dotted all over—it was a most beautiful picture, with the river winding through. There was a tree I had never seen before on its banks. It was the willow tree, called in these parts a Sally tree. It is quick growing, and ours were pollarded, i.e. cut off at about six feet from the ground, where they split into many small branches. At five or six year intervals these new branches were cut off when they were nine inches in diameter. They were cut into lengths of five feet, and then the hard part came; they were usually split into three-sided stakes about three inches thick, all done with an axe. The stakes were stored in an open-sided shed called a wainhouse to dry out, and all the small branches were tied into bundles for kindling. Nothing was wasted—even the chippings were bagged up. The Sally trees looked quite stark for a while but they soon sprouted new branches from their knobbly tops. The old trunks went rotten in the middle, so many of them had holes going up their full length. This was a favourite place for foxes to hide up in during the day. I often disturbed a fox in one, and it would jump into the river and swim across. On the other side it would look back in a most indignant way.

The Sally stakes were used for hedge laying, a winter job that I really enjoyed. In the border counties between Wales and England there is a lot of stock—cows, store cattle, fattening bullocks, sheep and lambs—and the fields have to be stock-proof. A farmer whose stock keep getting out is a bad neighbour, and a lot of damage is caused in a very short time, so everyone took a lot of trouble to keep the fences in good order.

With the land being so fertile, hedges grew fast and they had to be cut every year. A newly laid hedge would be cut both sides and on the top for four years. Then for the next three years it would only be cut up each side, allowing the top to grow high. After these seven years it would have become weak, with holes appearing at the bottom. It was now ready to be laid again; the thickness and length of the stems were just right for 'pleaching', which involved making a diagonal cut just above ground level so that the stem could be bent over and laid. This task had to be done when the sap had 'gone down', say from November to February, when the sap started to rise again. This fitted in well with the farm routine, because usually at this time all the cattle were in the yards so it was only the sheep that had to be moved.

The first job was to take down and roll up the old barbed wire and pull up the old posts. A lot of them would have rotted off at ground level, so they went into the cart to be sawn up for firewood. Now we could get at the hedge and pull out the old dead wood. Next we had to thin out the growing trees, because there would be too much bulk for a nice low new hedge. Some would therefore be cut out and laid on the ground, with the butts facing the hedge on both sides; it was much easier working if they were as tidy as possible. This cutting out was mostly done with a 'hacker', a small hand knife about the size of a butcher's cleaver. Only if the hedge had been left a good many years over the seven would an axe be used. We ended up with a thin, straggly line of tall trees.

Everything was now ready for the hedge-laying, and this was the part where each person could stamp his own mark on the landscape. As with stone walling, it was possible to tell who had laid the hedge by the style. And as with stone walling, I was not very good at it but I loved to do it. One year Tom and I did over a mile; it did look good and we were proud of it for the next three years! The aim was to make a hedge that would turn a 1700-pound bullock and also stop sheep and small lambs getting through. We took the tall young trees and made a long diagonal cut down the side as close to the ground as possible until we could bring it down and lay it at a slight angle along and through a line of stakes. These were the stakes cut from the Sally trees, dried and pointed. They were driven in along a line just to one side of the centre of the hedge. The distance between stakes was measured by placing our fingers on top of the last stake in the row; the next stake would go in so that its top was at elbow level. They were driven in at an angle, so that the top of one stake was directly over the bottom of the next one.

We were now ready to bring down the pleachers. We cut each one until it gently hinged down and threaded it through the stakes until it reached the top of the fourth stake. It was now at the correct angle and was quite firm, as it had been woven though four stakes. It did not have to go through every one—it may pass one, two or even three stakes before going through—but all the pleachers had to finish up on the same side, usually the opposite side to the man laying the hedge. So the new hedge had two distinct sides. The hedge-layer's side, which just had the trunk of the trees lying against the stake, was called the 'clean' side, and the other, which had all the small branches and the tops overhanging, was called the 'brash side'. As we progressed, the clean side would have six strong pleachers up each stake, with a few small branches that had been put in to fill any weak spots in the brash side. These small branches were cut from the lines of thinnings, pushed well into the soil and woven into the hedge in such a way as to make their bushy tops fill in any weak spots. Like the pleaches, they always finished on the brash side.

Every so often, one of the long pleaches would be allowed to run along the top to make a good solid top. During the hedge laying, the tops of the stakes protruded about six inches above the hedge. These were not driven home until all that length of hedge had been finished. It was possible then to make a good smooth line over the whole length by either leaving some a little proud or driving some in a little further so that it looked good to the eye. The other job that was done when the whole hedge was pleached was to trim the brash side into a nice straight line. We used to cut off the backs of the pleaches (that is, the bit of trunk that was left sticking up after the pleach had been laid) as we went along, but some people cut them off at the end as the hedge was much firmer to cut against. One final job was to go along with a spade and re-cut the bank and fill up any hollows along the base of the hedge. It was now time to clean up, cutting out any wood that was thick enough for firewood and burning the rest.

As most of the hedges were hawthorn, with a few hazels in them, we had to have strong, thick leather gloves. These were made from cowhide, and were actually just mittens with a thumb, and the fingers all together. A cap was essential, as was a thick jacket and tough boots. Even so we would often get badly scratched or have thorns stuck in all parts of our body. Despite this, the work gave me a great deal of pleasure.

Once the sap had started to rise in February, hedging had to stop, not only because it would harm the hedges but because it was coming up to one of the busiest times of the year: spring. As soon as possible all the grass fields had to be chain harrowed. One horse pulled a chain-link mat, nine feet by four feet with rows of spikes every nine inches, which ran over the surface so that the spikes tore out any grass that had died over the winter. This freshened the field and left it with a nice striped effect, a bit like when a lawn is mowed. This was the first of the big walking jobs, and my feet were usually a bit sore afterwards.

During the long winter nights, sitting in the dining room in the glow of a wood fire and oil lamp (for we did not have any electricity in Herefordshire either), Mr Downing told me how he worked on a seven year rotation. Starting with a field of grass, he would plough this up and plant winter wheat. The second year the wheat stubble would be ploughed and oats planted, the third year the oat stubble would be ploughed and a root crop, either sugar beet or mangolds, would be planted. The following year, after ploughing the ground the root crop had grown in, it would be planted with barley. When the barley was about four inches high (about May) he would spread grass seed amongst it so that when the barley was taken off, the grass would be established. For the next three years the field would be grass again, after which it would be ploughed out and planted with wheat, and the rotation would start all over again. Seven of his twelve fields were in different years of this rotation, another three were the river ground which was not ploughed, and two were 'passage fields', that is fields that had to be grass to allow us to move the stock to all parts of the farm. So now it was possible to work out what was going to be planted in each field each year. When, after the hedging and chain harrowing was done, I asked Mr Downing what the next job was—in what order the roots, oats, barley were to be planted—his reply was: "You have got to be in tune as well as in time, just like music." This meant that the seed bed had to be of the right texture as well as in the right time scale. If the field had been ploughed in winter and the frosts had weathered the ploughing so that a fine seed bed was produced, barley would be planted first. So, following the rotation, we would look at the field that had roots in the previous year. Oats will stand a rougher seed bed than barley, so the seed bed could be worked down with cultivators, without the help of the frosts. Oats were usually planted in March or April, so we looked at the field that had had wheat the previous year, and so on through the rotation, taking on one field at a time but making sure that the work was done at the right time to get the best yields.

TWO OR THREE TIMES a year during the winter we would have a day's threshing. It was an exciting time for several reasons. The night before, the big traction engine would draw into the rickyard, pulling the threshing machine with the wire baler and batten maker in a long line behind—a total of about twenty-five yards of slow-moving machinery. The steam engine was mostly fired by wood; coal was only used if the supply of wood ran out. It was possible to follow the route of the engine by the absence of any loose wood; every bit of timber that could be pulled out would have gone—old gate posts (and some that were only loose), hedging stakes and half-dead branches of big oak and ash trees would all disappear into the fire box of the huge engine as it plodded its way slowly around the farms.

But what a sight it was! Smoke bellowing from the chimney in front, steam hissing out all over its bulky frame and old Bob Kington at the ridiculously small steering wheel, with a big grin on his face and a cap like a railway man's but made out of leather, perched on his head. He smiled down from his high perch at the assembled crowd of Mr Downing, Tom, myself and perhaps three other casuals who had been drafted in for the big day, and he would ask which stack was to be done first. He would unhitch the baler and batten maker in the paddock, manoeuvre the threshing box into perfect position with the mighty, cumbersome engine, where chocks would hold it into position. He would fetch the baler, and with the same precision he would set it up dead in line, with its four-foot diameter drive wheel just the correct distance away from the wheel on the threshing drum that drove it. The engine now chugged back round to the other end and stationed itself so that its huge flywheel could drive the threshing box. The big heavy drive belts would be heaved on to the big drive wheels; they were always crossed when working. Whilst this was going on, a second man who had come with the machine would be unpacking the threshing box, folding out the platform where the sheaves would be received, putting up the low retaining sides on top and removing the tarpaulin that covered the threshing drum.

It was all go. The sacks for the corn were brought to the back of the threshing box. They were hired from the railway and were very thick jute; they would hold 280 pounds of wheat. They were lifted by a sack truck to a height of five feet, so they could be put across someone's shoulder and carried away, either up into the granary or under cover in the wainhouse, where they would be stacked two-high, ready for transporting. When filled, they were tied up with sisal (a very coarse string), with what we called a Lincolnshire potato knot, a type of clove-hitch.

A large light hessian sheet about twelve foot square was needed to carry away the chaff and pulse (broken wheat heads and small bits of straw) from the threshing box. This was usually blown through a large pipe a couple of yards from the machine, where the sheet was laid on the ground, and a large pile of chaff would be raked on to it. The corners were then pulled into the middle and, holding them in one hand, one of us would throw it on to our back and carry it away to the chaff house to be used later for feeding the cattle. This was the worst job of all. It was very dusty because we were working in a constant stream of blown chaff. Although the sack was very light, I felt a bit like an ant carrying a large leaf.

The baler at the other end of the threshing box was powered by another crossed belt. This drove a ram that pressed the straw into a very tight oblong block, which weighed about 110 pounds and was tied up with two lengths of wire. Each bale was pushed out by the one being made behind it. They were then carried away back under the barn and stacked where a stack of corn had been.

There were usually two men on the corn stack. One followed the courses of sheaves round and round, picking them up one at a time and passing them to the other, who was on the edge of the stack. He then passed them to the man on the threshing box, who cut the strings and fed them into the drum that threshed the corn out of them. This had to be a smooth operation, with all three working together. The man going round the stack had to know just where his next sheaf was, so that he could pass it to the second man just as he had delivered the previous one to the man on the box, who had to feed the drum evenly. He could not drop the whole sheaf straight in. If these three men worked like clockwork the threshing was done well and all the other people got into a good rhythm.

My job was usually corn carrying, and sometimes by the end of the day my legs were buckling a bit and my neck was sore from the rough sacks and dust, but after a shower and one of Mrs Downing's meals I felt great. This shower involved taking all my clothes off except my underpants and sitting under the pump outside the back door while someone pumped water over me; we did not have either mains water or electricity.

The threshing gang came into the house for supper. There would be about nine of us, and we exchanged local news and listened to tales of fording rivers and crossing wooden bridges with all the threshing tack. One of the accessories Bob carried on the box was a roll of half-inch chicken wire. He told us that this was for putting around the sack being threshed, in order to catch the rats! At one farm down by the river they had used it and caught 146 rats. Bob Kington was very fond of his steam engine, which was called Nulli Secundus ('Second to None'). He would have an oil can in one hand and a large ball of cotton rag in the other and would fuss around his engine all day long. He had another love, and that was ploughing in the local ploughing matches. He was so enthusiastic that I got the bug.

THE HARVESTING of the sugar beet was one job that I was pleased to see finished! We started just after Christmas, ploughing out the beet. One horse pulled a blade that went down the side of the sugar beet. This blade had a foot on it that lifted the soil under the beet and so loosened it. The beet can grow to about ten inches in diameter and are very firm in the ground. Without this ploughing out, it would be impossible to pull them up.

After ploughing about two acres, the next operation was to pull them up, one in each hand, and knock them together to shake all the soil off the roots. They were then laid down in a row with the roots all facing the same way. It was cold work. When the frost and even rain was on them, my fingers went blue and there was a lot of beating of hands to try and keep a bit warm.

The tops were then cut off with a sharp root knife. The usual way was to pick up the beet by the root and, holding it at arm's length, cut off the crown, complete with all the leaves. The root was then thrown on to a heap. All the beet in about a three-yard radius would make one heap, then another heap would be started about six yards from the first one, and so on down the length of the field. If it was frosty, the heaps would be covered with the tops to stop the roots freezing.

The crop was now ready to go to the factory to have the sugar taken from it. To provide an even flow to the factory, permits were issued throughout the season. When our first permit came through we had to remove the tops from the heaps (called tumps) and load the roots into a cart with a sugar beet fork—a wide fork with tines that are fairly close together. Each tine had a ball about the size of a marble on the end to stop it sticking into the beet. When the cart was full it was a two-mile journey down to Kingsland railway station, where we unloaded the beet by hand into a goods wagon. It took about four cartloads to fill a wagon. We grew about twelve acres of sugar beet, so it was quite a long job to get it all away, but doing the field in two-acre blocks, all in different stages of being harvested, we were able to vary the work, which helped a lot.

The other root crop we had to harvest was mangolds. This was done before Christmas and was a bit different, as the mangold grows above ground and the top leaves are screwed off by hand; they are not cut, as this would make the mangold rot. Mangolds are about twice as big as sugar beet, so we soon had a big heap in the rickyard for cattle and sheep feed in winter. They were not used before Christmas because they had to mature. We loaded them into the carts with a short hay fork; unlike the sugar beet, the tines stuck into the mangolds—like taking pickles out of a jar.

On the first threshing day, which was soon after harvest, the straw was not usually bailed. Instead, it went into the batten maker, which had some packing tines which lightly pressed the straw into a big cigar shape which was then tied around with a string about eighteen inches from each end. This was then built into a stack to be used later to cover the mangolds with. Only the wheat straw was done in this way, as it was the best straw for thatching both roots and hay or corn ricks. The straw was held down on to the mangold heaps with a few shovelfuls of soil, after which old hedge trimmings were piled up on top to protect the heap from frost.

WHEN THE WINTER PLOUGHING was done it was time to plant the winter wheat. Wheat followed grass in the rotation, so as soon as the ground was right, the old grass was ploughed out and immediately worked down until we had a good rough seedbed. Fresh ploughing did not always work the soil into a fine seedbed, but as this was not required for winter wheat it did not matter, so long as there was enough fine soil between the clods to cover the seed. The seed germinated and after the winter frosts the field was harrowed with a set of very light harrows, to break down the clods without harming the young growing wheat. The wheat, which was drilled in seven-inch rows, now grew away quickly and became 'shut drill', which means the rows spread out and met so that all the soil was covered. The sooner this happened the better, as a cover stopped the weeds from growing and we would have a cleaner crop. It also cut down the time spent on that most boring job, thistle spudding. The wheat was then left to grow and ripen in late August.

The first combine harvesters were just coming into the country from America at about this time, but we had yet to see one. Our corn was cut with a binder, but first we had to cut round the outside of the field with a scythe. This was called 'hooking and crooking'. The loose-cut wheat was pulled into sheaves with the crook, then we took off a few strands of wheat to make a 'straw string' to tie it up. This made a clear 'roadway' around the crop for the horse and binder. When about six rounds had been cut with the binder we would start to put the sheaves into stooks. We stood two sheaves up, leaning them together with their heads up, repeating the process on each side of the first two, making a stook of six sheaves. Sometimes, in a heavy crop, eight sheaves would be put into a stook. These stooks were then left in the field for three Sundays to dry out and ripen. The wheat had to be cut when it was not quite ripe, so that it would not fall out of the head when being handled. The stooks would turn the rain and keep the wheat dry during the ripening period.

When the binder had cut about three-quarters of the field, we would take a break from stooking to chase the rabbits that were now starting to leave the ever-decreasing standing crop. It was a hectic half hour. The rabbits were confused by the crop being cut, and the sheaves that were lying on the field were like an obstacle course for them. The rabbits kept running into them, which gave us a chance to catch them. A good many escaped but we would usually finish up with thirty or forty. Unfortunately these were taken by Mr Downing, so Tom and I did not benefit from them, but it was a bit of good fun and light relief from the work.

The wheat, oats and barley were all cut and stooked in this way. Barley was the hardest to stook, as the sheaves were so short and the long awns (whiskers) on the grain were very abrasive and itchy. After the third Sunday we would start to haul the stooks into the Dutch Barn in the farmyard with a horse and dray. Tom would stack the sheaves on the dray as I pitched them up to him two at a time. When I could not reach any higher we would go up to the farm and unload. We had three drays and we would fill all of them at night, leaving them to be unloaded first thing in the morning. This saved time because we had to wait until the dew had dried off the stooks in the morning before we could start to load. Mr Downing would build the stack, Tom would unload the dray, passing the sheaves to me, and I had to pass them to Mr Downing head first; woe betide me if one went butt first.

Mr Downing was a very good stack builder. Sometimes a corn rick had to be built outside when the barn was full, and this was when he was at his best. It was quite a job, as first of all he had to estimate how big to make the base. It was no good having it so big that there were not enough sheaves to finish off the roof, but equally it was no use having half a load of sheaves left over. A well-built rick was a work of art. The corners were rounded and there was an outward slope from the base to the eaves, then the roof sloped in from the sides and ends. It also had round corners. If it had to stand for a few months, it had to be thatched, which was also an art. After each two or three loads had been unloaded on to the rick, Mr Downing would get a shovel and go round and round it, knocking in any sheaves that were slightly out of line. It was an art I never mastered. I remember Tom's comment when I finished my first small rick: "It looks like a sinking battleship." He could build a good rick, but he was much better at building loads of sheaves on the drays—he never had one slip off and, believe me, most people lose one or two during a harvest. On one occasion the boss built a rick on a cartwheel. A cartwheel is not flat; the hub in the centre is concave. He laid it on the ground so that the concave side was facing up and then drove a stake through the hub into the ground until the top was level with the top of the hub. The wheel was about five feet in diameter and could now be turned, as the hub was lower than the rim and it was therefore clear of the ground. He then started to build his stack, going round and round. It was a perfectly round stack, and when it was finished it was about fourteen feet high and could still be pushed round. I have never seen it done anywhere else—it was wonderful.

When all the harvest was in under cover, it was a time of great relief and a sense of achievement. The whole year seemed to climax at this point. The farming calendar was programmed to bring together all the different forms of agriculture, whether stock or arable farmer, to a common meeting point. Harvest Home or Harvest Festival is a very old celebration, with great meaning for the farming community. Even a sceptic like myself found the Church's recognition of this event heart-warming, and I loved to sing "We plough the fields and scatter" with great gusto.

MY FIRST ATTEMPT at ploughing with horses was not very successful. Mr Downing sent me out with Tom for a couple of days to learn how to do it. A horse plough turns the ground one way only, from left to right, so to start we placed a line of posts down the field at about twenty-yard intervals in a dead straight line. We had to ensure that they were very straight. The three posts farthest away from where I was to start were close together, being about twenty, then ten, then five yards apart, so that it was easier to stay straight right to the very edge of the field. Now all I had to do was to lift the wheels of the plough so that the share (point) went into the soil to about four inches deep. This share was followed by the 'breast', which lifted and inverted the furrow as I progressed down the field. If I kept the two horses dead in line with the stakes, I should have a nice, neat furrow stretching out behind me, straight as a gun barrel. My first attempt was more like a snake! Apart from looking awful, this made it difficult when I had to turn round and come the other way down, turning a furrow that exactly met with the first one.

The first furrow up and the first furrow down together are call the 'cop' in Herefordshire, and a 'rigg' in Yorkshire. The idea is to continue ploughing on either side of the cop (called 'gathering'), until thirty yards has been ploughed (fifteen yards each side of the cop). Then a new cop is made sixty yards from the first one, and parallel to it. I gathered around this one until another thirty yards had been ploughed. This left thirty yards of unploughed land between the two blocks. I now ploughed up the second block and down the first, gradually closing the gap between the two, called 'slitting', until they met. If the ploughing is straight, this meeting will come together perfectly.

That is the basic art of ploughing, but to do it really well you not only have to keep straight but also keep a constant depth and turn the soil over cleanly so that no straw stubble is showing. The soil should also be firmly pressed so that the ploughing 'walks well' and is not 'fluffy'. The cop should be the same height as the surrounding furrows (not easy) and the finish should be shallow so as not to be too rough for the following cultivations. Ploughing like this is not only a joy and a pleasure to do but also leaves one with great satisfaction. I was later to try my hand at 'match ploughing'.

Ploughing with horses started early, as I first had to catch the horses, take them to the stable and give them a rack full

of hay, then curry comb and brush them down. I would then
have breakfast, after which I would harness them with the
long traces and walk down to the field (we always walked,
never rode), hook up to the plough and plough until bait
time, 11 o'clock. Bait time was when we had a break for a
flask of tea and a sandwich. Tom and I would always try to
have bait together to catch up with the news or gossip. The
horses would have a nosebag put on with a few rolled oats
in, but it was not a long break (just fifteen minutes) because
both we and the horses would have been sweating and we did
not want to get cold. Then away we went again until lunch.
The boss would come down with a change of horses for Tom.

I would unhitch and bring my team and Tom's first team
home whilst Tom carried on until teatime. After I had stabled, fed and watered the horses I would have my work to do
feeding and bedding down the Hereford cows, calves and
bullocks and taking a walk around the sheep.

One day when I was ploughing, it was going quite
well when suddenly I felt a tremendous thump, after which
I remember nothing. I came to under the hedge on the

headland (where the horses turned at the end of the furrow), with a pain all along my back and chest. I had no idea what had happened. It turned out that the point of the plough had struck a large stone deep in the soil, the horses had given an extra pull on feeling this obstruction and that had suddenly and with great force lifted the handles of the plough, which had hit me under both armpits and thrown me up into the air, knocking me out! I recovered and was able to continue ploughing, but from that time on I was nervous and tended to walk a little too far back from the end of the handles, which did not help me to plough well. But I have always been lucky in my life. It was only about a month after this that Mr Downing bought a standard Fordson tractor, and I was given the job of driving it as Tom loved his horses and did not want one of "them new-fangled things". It was far too noisy and smelly for him, and besides, "folks would think he was daft if he talked to it like he talked to his horses." So now I was stock man and tractor man.

I really did like ploughing with the tractor. The son of the farmer next door, George Young, had been a very good ploughman but had given it up to concentrate on his new hobby, welding. However, he said he would come round and teach me to match plough. Like everything, to do it well needed practice, practice, practice, and this I did night after night. During the day I had to get on with my day's work, but after tea I would go out and learn the finer points of how to keep not just straight but dead straight. My plough turned two furrows over each time, and they both had to be exactly alike, well turned over, clean, with no old straw showing and firmly pressed to make a good seed bed for the next crop. It sounds quite easy, but in fact it is very difficult.

The main difficulty is to get all the furrows matching, like peas in a pod. It is quite easy to match the two that the plough is making, but it's the next two that are difficult. If the plough is not exactly right, these two will still match each other, but there will be a slight difference between them and the previous two. The ploughing will show good furrows, but in doubles. It will be plain to see the first and second furrows in each bout—called coupling. It should look uniform and it should be impossible to tell which furrows were drawn out by the plough. It takes many hours to get the plough set right and while this is being done, a straight line must be maintained.

After weeks of practice, I entered the Leominster ploughing match; I was petrified. The date of the match was 8th October 1949. By the end of September I had decided I was not going—it was too much of a strain on Mr Downing, George and myself, and Tom said that I was just daft and wasting a lot of good time. However, after a sleepless night on the 7th October, I was up early, fed the cattle and sheep and set off to the match. It was a good job I arrived early, because the field we were to compete in had been covered in muck from the cattle yards, and it was drizzling. My tractor had rubber tyres, and the combination of rain and cow muck meant that it was very slippery. George had followed me in his van, and when he saw the conditions he said that we must change the tyres for iron spade-luggs, wheels of steel with flat pointed studs in the rim. He went back home to fetch a pair whilst I jacked the tractor up and took the rubber tyres off.

By the time I had finished it was time to draw my plot number out of a hat. We were allowed four and a half hours to plough one and a third acres. There were twelve ploughmen in my class, which was for 'Agricultural Employees'. I noticed that some of the more experienced ploughmen had brought a spade-lugg with them, and I was surprised to hear that they had been in the day before to see where the field was and what had to be ploughed. I stored this in my memory for the future, but as I only had a bike I knew would not be able to travel far. Moreover, I had plenty of other work to do for Mr Downing. It turned out that many of the ploughmen worked for bosses that wanted them to win ploughing matches. They were all very competitive.

Starting time came and we all started to put up the 'copping sticks' for a good straight start. I got mine up, but my tractor was still up on the jacks, George had not arrived back! Fifteen minutes later I still had not started, and I was just about to put the rubber tyres back on when George tore into the field with the spade-luggs. They weighed about 220 pounds each, so it was another fifteen minutes before, with the help of the stewards, I was able at last to get going. I thought that at that point it was too late for me to do any good, so I decided just to enjoy myself, have a good day and learn as much as I could about the competition. It soon became apparent that I had an advantage, as I had two spade-luggs. The other ploughmen who had brought spade-luggs had only brought one, for the side that ran on the field; they still had a rubber tyre on the wheel that ran in the furrow, so they were getting a little slip. Their ploughs were therefore not as steady as mine, and my work was that little bit better. The men who were on rubber tyres only were really struggling. I also had another advantage in George Young. He had been a champion ploughman, and although he was not allowed in any way to assist me his consistent encouragement made me determined not to let him down. I had also got off to a very good start and, although I was still half an hour behind the rest of the competitors, I was slowly catching them as a result of not having any wheel slip.

At eleven o'clock 'bait' came round—a bottle of beer and a sandwich. Most men stopped for this, but I took my sandwich and kept going. George was delighted that I did not drink beer; he said his voice needed the extra lubrication because of his constant chivvying, so he had mine. I wanted to catch up as soon a possible because towards the finish there is a lot of measuring to do, to make sure that the two sides are exactly parallel. So far my work looked fine, but of course I had not seen the work of any of my fellow competitors.

With about eighteen yards to go to the finish, my work was six inches wider at one end that the other. This was acceptable, but it had to be put right without putting a bend in the ploughing. I knew which way I would be going for my final furrow, so that side had to be kept dead straight. That meant that my six inches had to come off the other side. With a total width left of eighteen yards, and with one side being kept dead straight, it meant that I had nine yards in which to lose the six inches. My furrows were each ten inches wide, so I could take three eighths of an inch extra off the wide end each time I went down, and I would finish up just where I wanted to be. The concentration needed to do this without putting a slight curve in the furrow was tremendous. The plough was making two furrows, and it was impossible to alter the width of the second, so all the three eighths had to be put on to the first. But this had to be carefully brought back to normal width by the time I had travelled halfway down the stretch. However, any increase in the width of the front furrow would show in the finished work, so to counter this I had to make the front furrow shallower by winding the point of the body out. This of course had to be slowly wound back again as the furrow width was reduced. If that sounds complicated, well it was!

As I was working away at this, I knew I was still behind on time. At last I got down to the last four furrows and I was pleased to find that it had worked out all right. Then I looked up to see George standing at the end of the furrow with a crowd of people around him. I looked at the other end and there was another crowd of people. Where had they come from? I looked for the other competitors, but they had all finished. Oh no! Was I out of time? I started the long crawl down to George's end, drawing two shallow furrows so that my finish did not look too deep. George looked worried, so I asked if things were all right. "Yes," he said, "grand," but told me that I only had five minutes to finishing time. I knew then that I could not finish as the plough had to have various adjustments made before my final run down.

The stewards allowed George to help me alter the plough, and whilst we were doing so he said, "Take your time, don't worry if you overrun the time, just make a good finish." Once again that was easier said than done; when I was ready to set off I could see the judges just starting to make their final walk over the twelve plots (they walk up and down three or four times during the day to see the progress). It was a nerve-racking time; the slower I went, the better the work would be, but I did not want to miss out after gaining so well during the day. I was twenty yards from the end when the judges arrived, and they were still standing by my last furrow when at long last I finished. Two minutes later they were still there. Then they moved off over the last three plots. George wandered down and said that I had not done badly for the first time out. I was just drained. I was suddenly so tired. I never even walked back along my ploughing to see what it looked like. I unhitched the plough, and we fetched the jacks and replaced the spade-luggs with the rubbers.

We were just removing the jacks when I noticed that the stewards were putting pegs with cards on into the ploughing. One stopped at my plot, bent down and placed a peg on my cop (start). I looked at George, but he carried on working so I did as well. Then he said, "It looks as though you've got first prize—aren't you going to pick it up?" I couldn't believe it—not me! I went to look, and yes, it was a red card! Suddenly I wasn't tired any more, I had to go and look at my ploughing, then at the other eleven plots. Every one of the other competitors came and shook my hand. Some of these men had been ploughing for many years and were previous champions, but they all had big smiles on their faces, and I could tell that they were genuinely pleased that I had joined their ranks. It was a most wonderful moment, and today as I sit here with that first ploughing cup within reach, it brings back those happy times and also the memory of some very special men.

MR DOWNING HAD his own herd of pedigree Hereford cattle: the Moss Hill herd of about twenty cows and a bull. To be in the top flight of the pedigree world required a lot of showing, and a good sprinkling of prizes won. We did not do this, but nevertheless we had a good reputation for producing bulls, often for crossing with Friesian dairy cows to produce cross calves of a high standard for the butcher. We had a few dairy farmers that came regularly for our young bulls. These were always given extra feed and were well groomed and turned out well. They were a picture: deep red and pure white with their coats shining (we use to feed them linseed to produce that effect) and well curry combed and brushed.

They were given regular doses of medicine to control worms. If this was not done, the cattle would look unhealthy. This was called 'drenching', and was usually done by putting the medicine into an old cow's horn, then pouring it down the throat. Sometimes they would cough just as the medicine was going down—you can imagine the result! We used to wash them with soap and their tails were fluffed at the tips by taking a few end strands in one hand, and with the thumbnail of the other pushing up the strands. This would make the last eighteen inches of the tail bush out to about four times its normal size.

They were taken out to drink at the water trough each day on a halter to get them used to being led. When a buyer came along, the bull would have a new white halter on and I would walk it round the fold yard, whilst Mr Downing and the buyer stood in the middle. These well-turned-out, good-looking bulls that walked well on the halter (led by a smart young lad in a pristine new white stock smock) commanded a premium price. It was well worth the days of being dragged around the yard with me on the halter and the boss hanging on to the tail, with both of us finishing up covered in muck.

We were proud of our Herefords, especially when we sold one to the Milk Marketing Board. The average price for our bulls was about £120, but the Milk Board paid us £600! It was a good day. We would only sell five or six bulls, as of course about half of the calves would be female, and only a very few of these would be required for herd replacements. So we would have two yards in winter, one for heifers and one for bullocks—the castrated bull calves that were not good enough for sale as bulls. We also had a few extra bullocks, as some of the cows had far too much milk for one calf and Mr Downing would buy bull calves to use this surplus milk. Sometimes a good cow would rear her own calf and an orphan together, and when they were weaned she would have a third calf put on her, so we always had two full yards of young stock throughout the winter. All these yarded cattle were fed on home-grown food. This was wheat chaff (the coating around the wheat grain), mangolds put through the pulper, and barley ground on the farm. These were mixed daily with a shovel throughout the winter and carried in big baskets to the stockyards.

In April the bullocks would be taken to Leominster Market, right in the centre of the town. This was one of the most exciting days of the year because we used to walk them there, about five miles down the main road! When the bullocks were first let out after being shut up in a yard all winter, they just went wild, running round and round the field jumping and snorting. When they had got over this first taste of freedom and settled down a bit, Mr Downing would say, "Right, let's go." I was surprised to discover that this just meant Tom, myself and an extra lad called Bob, specially drafted in as a drover. The boss said he would catch us up later in the car!

The road gate was opened, which was obviously the cue for the cattle to have another mad fit, but this time down the road. Tom and I were in front and Bob was behind with Bob the sheepdog. The plan was for Tom to hold up the cattle whilst I shut any open gates. We didn't get off to a very good start. Opposite the very first open gate was another. "Snap," shouted Tom. I chose the wrong one first. It was off its hinges and took an age to close, by which time the bullocks had out-run Tom and were now charging around a twelve-acre field. Bob the sheepdog was in hot pursuit, but just seemed to encourage our mob to greater frolicking. "You go in and help Bob," said Tom. I went in, shouting at the sheep-dog to stand still so that the cattle would settle down. This worked, and after a while we had them back out on the road; the only thing was that I was with both Bobs at the back of the herd with only Tom in front. "Send Bob up to me," said Tom. "Which Bob do you want?" I asked. At this Tom flipped. I had never heard him swear so much before—quite unreasonable, I felt. However, it appeared that we were approaching a crossroads, so wherever Tom stood the bullocks had two choices of route: one right, and the other definitely wrong. Tom took the left hand side, the mob charged straight on—luckily, the correct way. The only thing was that Tom had now joined us at the back. The two Bobs, Tom and I all behind the bullocks, with no one in the front—and we had not yet travelled a mile!

We were almost through the village of Monkland, where the road goes over the River Arrow. On the right-hand side was the church. It had only a small gate but it was open, so in went one bullock. The others looked, then all decided to follow, trying to get through one three-foot-wide gate. Two got stuck, and while the others stood bellowing at the one inside the churchyard, Tom and I were able to regain our position in front of the herd. Bob (the man) moved the two beasts that were stuck in the gateway and we called the cattle over the bridge. The one in the churchyard came running after us, thinking it was being left behind. So there we were, all in order, with a long straight bit of road in front of us, and with no open gates as far as we could see. We did have one fright as a herd of cows in a roadside field charged up to the hedge to investigate our bullocks. We did think they might jump it, but it was a good, sound hawthorn hedge and it turned them. The bullocks were going well now, as they were running out of steam, having used up a lot of energy early on.

We were getting tired, not only from the running about but also the tension and responsibility. As we approached Leominster, the cattle actually had to be pushed along as they tried to eat the roadside grass. It was at this point that Mr Downing's Ford 8 came into view. As he passed us he said, "I see you're having an easy walk." I looked at Tom, who had gone red in the face but said nothing. As we walked into the centre of the town we could hear Mr Downing shouting instructions out of the car window: "Shut that gate," or "Just stand in that gap until we are passed" to various people, so he did help for the last half mile!

There was plenty of help in the market to pen and sort the bullocks into two even groups, and a market man came to wash and spruce them up. He did a fine job in a lot less time than it would have taken me. Those market men looked rough, but they were very good at their work and knew how to make any beast look its best. When the bullocks were sold they made top price that day. I said to the boss that I was pleased he had been lucky enough to get top price. He said that that it wasn't luck to get top price, it was hard work and knowing what you are doing, but he was pleased that I was pleased. Moreover, he topped the market with his stock most years.

DURING THE WINTER I was able to start catching rabbits again. I was still earning ten shillings (50p) a week, but Mrs Downing would not do my washing, so a lady down on Monkland Common did it for me. My usual Sunday morning job was to take it to be done. This cost me 2s 7d (13p) a week, which left me with just over seven shillings (37p) for the rest of the week. I had started to take notice of a lovely girl who lived on the next farm, but I could not ask her out, as I was so short of money. By now I also wanted to have my own small farm, so every penny I could bank I did. I asked Mr Downing for a rise, but he said I could not have one as he was still teaching me a lot of new skills—which I had to admit was correct. However, I had noticed that a rabbit would fetch one shilling (5p) down here. Knowing that rabbits were quite a pest on the farm, I asked if I could have the ones I caught. I was over the moon when he said yes. I had snared rabbits in Kettlewell, but here in Herefordshire there was stock in most fields, so snares were out. This left ferreting. I had never done this but Tom had, so I asked him if he would go halves with me at ferreting, and he agreed. For the next two weeks all my spare time was spent making hutches and going around looking at ferrets, handling ferrets and buying ferrets. Our biggest outlay was for purse nets, which are placed over the rabbit holes before the ferrets are put in. The last thing I made was a box to carry the ferrets around the farm. Money was short, so we bought just three ferrets, two jills (females) and one hob (male), on condition that we could try them out first.

The first weekend out we caught twelve rabbits. If we could do that again the following week, the ferrets would be paid for. We caught eight, and spent the Sunday afternoon digging a ferret out. What had happened was that one of the jills had killed a rabbit down the burrow, eaten its fill and gone off to sleep. It is for just such a situation that we had a hob. We fastened a collar on him, with a long piece of line fastened to it. We put him the rabbit hole and fed the line in. The line had a knot at a yard, two knots at two yards, and so on. The hob went to the jill, so we were able to work out roughly how far in they were. Unfortunately, rabbit holes do not go in straight lines or keep to an even depth, but we could follow the general line. They were quite deep, so it took quite a long time, but what a fine start—two weeks and only four shillings (20p) left to pay off on the ferrets.

Ferreting is a skilled job. Most warrens run along hedgerows, with holes on both sides of the hedge. If they were out in the middle of the field they would be disturbed by activities such as ploughing, so the warrens out in the open are on rough land that has never been cultivated.

We would start at one o'clock on Saturday. We would take a graft (a long narrow spade about five inches wide), two light hedging bills (long-handled knives), purse nets and the transport box with the ferrets in it. The ferrets would not have been fed since Friday lunchtime, so they would be a bit peckish. We would approach the chosen hedge with the wind in our faces; we always worked into the wind so that our scent was not carried forward to give the rabbits warning. I would be on one side of the hedge, and Tom on the other. From then on we did not speak. We very quietly cleared any brambles or small bushes away from the holes with the hedge bills. All the bits of cut-off trash were cleared well away from the hole so that it would not stop the purse net closing and enable the rabbit to get away.

When all the holes had been cleared, we placed the nets over them like a veil and held them with a peg pushed into the ground. A cord from this peg went around the circumference of the net so that when a rabbit bolted from the hole it hit the middle of the net, which was carried forward and enveloped it. The peg held the cord fast so that the net was closed and the rabbit could not get out. It was then just a matter of holding the rabbit by its back legs, removing it from the net and killing it with a sharp blow to the back of the neck. The net was then replaced over the hole, and it was all done as quickly and quietly as possible.

Normally the rabbits would bolt from the hole as soon as the ferret was put in, as they could smell it and would not have to be chased out by it. But if we made too much noise they might prefer not to come out to face us, so it was vital to keep as still and quiet as possible. The ferret would soon negotiate all the passages underground and if they were empty, she would come out. It happened sometimes that we spent a lot of time netting up all the holes, and then the ferret would not go down as she knew that the warren was empty—no one at home. If this happened we would always try the other jill, but she would very seldom go down. The hob was never used to bolt rabbits; he was a big fellow and was quite capable of catching and killing a rabbit in the burrow. He was therefore only used to find a lost jill.

Most warrens had special boltholes, usually two or three yards out in the field. They had no piles of soil outside them; they were just very small holes about four inches in diameter, often in longish grass. These holes also had to be found and netted. One of these would often be missed, and a rabbit would suddenly appear as if from nowhere and he would have the last laugh. We became quite expert at finding them, however, and once they had all been netted, Tom would put his thumb up and the ferret would go in. When the warren was empty we would hang any rabbits we had caught on the barbed wire fence to be collected on our way home, pick up the nets and move on to the next warren.

This was what we did every Saturday and Sunday throughout the winter. Most of the time it became quite routine, but sometimes we had a bit of luck and caught two rabbits in one net. Once, in the two years we were there, we had three in a net. Luckily this happened right at Tom's feet and he was able to grab the third one, as it was only half in the net.

The best catch we had was a great surprise. The ferret had gone in, but nothing came out, and all went quiet. We put the hob in and it stopped four yards in—it looked like being a long day. Tom was standing about two yards from where the warren started, tight under the hedge. This was our usual position, for if we were in front of the holes the rabbits would come to the entrance, see us, and turn back to face the ferret. Tom said that thumping had been going on under his feet, so he suggested digging there before following the line. He sank the graft in and it went straight into a burrow about five inches below the ground. When he removed it, out popped the jill ferret, its head covered in blood. Tom must have nearly cut it in two with the spade. All was well, however, as Tom thrust his hand in and pulled out the dead rabbit with the hob hanging on to it. So we now had both ferrets and the dead rabbit, and it had taken just a few minutes. Then Tom said, "Just a minute, there's another rabbit down here." He quickly rammed the spade into the hole to stop the rabbit escaping, and pulled it out, then another, and another. We couldn't believe it when we pulled out seven rabbits. They had run down a dead end and were boxed in by the dead rabbit and ferret. Unfortunately this only happened once.

I took all the rabbits to the butcher at Kingsland on my bike, slung over the handlebars and along the cross-bar. This could sometimes be a long job, as I was unable to peddle so I could only freewheel downhill, and had to walk on the flat. Once I was freewheeling down the road with a big load, when one rabbit got caught in the front wheel and I went head over heels in the middle of the road, rabbits everywhere.

We were soon catching about fifteen or twenty couples of rabbits each week, so we were making quite a difference to the rabbit population on the farm. It was also making a difference to my wages, and I well remember the day when my bankbook showed a balance of £100. I was so excited and my mind started to think about getting a smallholding of my own. Land to rent was £3–£4 per acre in Herefordshire in those days, so I had enough money for a twenty-five acre holding. Mr Downing was also letting me keep four ewes with his flock, so I was feeling good and really enjoying life. Many things were to happen, however, before I got my smallholding.

SEPTEMBER AND MARCH were the times when the rent had to be paid. This was the reason we usually had an early thresh-ing day: we used to thresh as much wheat as was needed to pay the rent. The rent was very important because, although tenant farmers' tenancy agreements gave them a lot of protection by law, they could lose all their rights by failing to pay their rent on time. They could, after a very short time, lose their farms, so this was the one bill that just had to be paid on rent day. It was quite an event at Burton Court. All the tenants were there at the same time. Mr Lane, the landlady's agent, stood behind the rent table in the great hall with his ledger and carefully, in copperplate handwriting, recorded the transaction. The great hall was a magnificent room with a huge carved black oak fireplace at one end and high timbered beams in the roof. After paying the rent, the tenants were invited to have supper, a buffet meal with lots of food and no restrictions on the amount that could be eaten. Some of the farmers had fine appetites. The second year that I was with Mr Downing he took me with him. It was yet another day that I will never forget; it made me wish that I could be a tenant.

Soon after this that I got on my bike without telling anyone, cycled down to Burton Court, knocked on the door and asked to see Mrs Clewes. I was shown into the hall, where Mrs Clewes met me. She was a lovely lady, very elegant, but I did feel very much out of my depth—as indeed I was. But I stuck to my guns and asked her if it would be possible to become one of her tenants if a small farm became available. She asked me where I came from. I said Moss Hill, and she said Mr Downing was one of her best tenants. But she left the letting of all the farms to Mr Lane, so I should go and see him. I did so a few days later, but unfortunately I never managed to impress him.

October that year was a month for decisions. After winning the Leominster ploughing match, I went to the dinner to receive my cup and was approached by Mr Tom Rogers of Broadward Hall, Leominster. He asked me to go and work for him to plough all his roadside fields. I thanked him but said that I already had a good job, and in any case, as I lived with Mr Downing I would have nowhere to live. "That's no problem," said Mr Rogers, "You can live with my farm manager at Corn Hill Cop near Leominster and I will pay you seven pounds a week." Seven pounds! Seven pounds! I was still getting ten shillings a week from Mr Downing and a farm worker's wage at that time was £3 15s (£3.75) a week. "You've got a new man," I said without any hesitation or thought of what I would say to Mr Downing.

When I gave him my notice the following morning he was not very pleased. I felt uneasy about it as well. Mr Downing had not only taught me a lot of new skills, he had also welcomed me into his home, so the following week was quite strained. During that time Mr Downing did offer to pay me a full farm worker's pay, and pointed out that I would have my keep as well. But there was still a difference of over three pounds a week, and in any case it was time to move on. I had another thought at the back of my mind. Mr Rogers had a herd of milking cows, and I reasoned that if I was to get a small farm of my own I would have to produce milk, as this was the only enterprise that would be viable on a small acreage. As I had only milked two or three cows by hand at Kettlewell, it would be a good chance to learn not only machine milking but also the day-to-day management of a dairy herd.

Then something happened to give me much more heartache. I received a letter from Mrs Middlemiss to say that Matt had developed arthritis and could no longer shepherd on Great Whernside. She asked me to go back to Kettlewell to take over the running of the farm. Matt offered to give me a half share of the profit, and I would also be the Moor Shepherd on Great Whernside. What on earth should I do? I was so much enjoying farming in my adopted county, I had hopes of acquiring a small farm, I still had a girlfriend and I had just been offered a new job with unheard of wages, but Sunter's Garth still held a large part of my heart. I badly missed the sheep and the moor, and I hated to think of Mr Middlemiss being restricted by ill health. One minute I had decided I was going back, the next I was staying. The thing that made up my mind was something I remembered my father saying many years before: "Never go back, lad, always try to keep going forward." So with great reluctance

I replied in the negative to Mr Middlemiss's offer. When the letter had gone, however, I was in quite a bad state for a long time. Indeed as I write now, forty-five years later, I still wonder.

My proposed new living accommodation was only three and a half miles away from Moss Hill, on the main road to Leominster. Mr Sanders was to be my immediate boss, and I had been told to see Mrs Sanders about my accommodation. As I pedalled up the long drive to Corn Hill Cop I wondered what my new job would be like, whether I would like the people and the different way of farming. I had noticed the farm during the previous two years, and I could tell that it was farmed very well—actually it was their herd of cows that charged up to the roadside when we were droving the bullocks to Leominster market—but it was to be the first time I had not lived with the farmer himself and the first time that I would not be a stockman. There was a flock of sheep at Corn Hill, but I was now a tractor man and would only look after stock as a secondary occupation. I began to have doubts about this, and when I arrived at the house I was very nervous. Mrs Sanders opened the door and I explained that Mr Rogers had sent me. She invited me into a most wonderful kitchen, with a huge Aga, a warm welcome, and a big cup of tea. A few weeks after moving in she told me that Mr Rogers had told her that he might be getting another workman, but never mentioned that I would require board and lodging. She said that it was on the tip of her tongue to say that I could not live with them, but fortunately she agreed to take me in. It was a wonderful home, and I was looked after like a king. For the first time I felt like an equal with the other staff.

I was very happy and thankful to be working and living with such lovely people. Mr and Mrs Sanders had three children: Bryn, Christine and Glenys. This was the first time I had lived with children about, and it was strange, but as all three were very active and were all learning to play the piano, there was never a dull moment. Their enjoyment of life was infectious, and these youngsters, without knowing it, showed me another side of life. I was happy throughout my time at Corn Hill Cop. One thing I did find awkward was that Mr Sanders asked me to call him Ralph, and his wife Jean. This was not easy as it seemed a little disrespectful, especially when I found out that Jean was Mr Rogers' sister. Mr Rogers used to ride around Corn Hill two or three times a week but he had five hundred acres to manage at Broadwood Hall where he lived, so Ralph was in charge at Corn Hill. It was good to see Mr Rogers and he always made a point of talking to all his staff whenever he came around.

I started to learn about the milking herd. After ploughing for five and a half days, I would spend Saturday afternoon and Sunday being nursemaid to the Ayrshire herd. I learned that the price of milk was much higher in winter than in summer, so our idea was to calve the cows in autumn so that the bulk of the milk was produced in winter. We would put a lot of effort into making good hay and silage in summer so as to get good yields of milk. The cows were milked in a huge cowshed, with the milking machines moving from cow to cow along its length. The machines were plugged into a pipe that went the whole length of the shed and continued on through into the dairy, where the milk was slowly run over a cooler to bring it down to a very cold temperature. After going through a filter, it was measured into ten-gallon milk churns ready for the milk lorry to pick it up and deliver it to Cadbury's dairy. At the dairy it was converted into condensed milk for Cadbury's chocolate.

I enjoyed working with the cows and was determined to learn all I could in case the day ever came when I had that small farm of my own. For Christmas, Jean and Ralph gave me a book called *The Principles and Practice of Dairy*

Farming by Kenneth Russell. He was Principal of Askern Bryan Agricultural College in York, and I don't believe anyone could have read it more thoroughly than I did. I could almost quote several chapters by heart. I would have loved to have gone to agricultural college, but I was getting practical experience and I was with people who really knew how to manage a dairy herd well.

The next twelve months flew by. The majority of my year was spent ploughing (Mr Rogers had bought me a brand new plough). I could now also be left in charge to do the milking. Hedge laying and harvesting were done by much larger gangs of men, and was much easier. By now I had managed to save £500 and had put my name down for one of the county council smallholdings. I had also applied for two others, but with no luck. Then in September 1950 I was ploughing a roadside field when Mr Rogers pulled into it and came across to me.

"How's it going?" he said. "Fine," I answered. "I thought so. I could hear you singing half a mile away." I did not realize that my awful voice could be heard above the noise of the tractor. His next question took me by surprise. "Are you still trying to get a smallholding?" "Yes," I said. "How on earth did you know?"

"I'm on the committee for letting the County Council smallholdings."

"Oh great, have I got one?"

"No, and you never will. You have got to have £5,000 available before you are considered."

"Five thousand pounds!" The Chairman of ICI at that time was earning £2,000 a year, so what chance did a farm worker on £350 a year have? Mr Rogers explained that it was not actually necessary to have the money itself, but one had to have the collateral to cover that amount. Unfortunately, in my case, it amounted to the same thing. If I saved every penny I earned it would be years and years before I got anywhere near that figure.

Mr Rogers could see that I was bitterly disappointed, and hastened to say that he had not come just to tell me this bad news, but to offer me a smallholding that he had just bought—if I would like it. If I would like it! What was he saying? I went from despair to elation in seconds. "Yes, I would love to have it."

"Wait a minute," Mr Rogers said. "You don't know where it is or how big it is."

"I don't care, if you are offering it I will take it."

"No," he said. "Go and look at it and let me know if it's all right. If it is we will talk about rent."

I would have none of this, however, and repeated that I would take it. So he held out his hand, which I grasped with most heartfelt thanks. I knew that once hands had been shaken a deal had been made—it would not be broken after that.

THE HOLDING, called Buckfield Farm, was right on the out-skirts of Leominster. It was 32 acres, 16 of which were an apple orchard. The rent was to be £7 an acre, £224 a year, to be paid half in March and the other half in September. It had a lovely small house, one of the rooms of which was completely round, as in years gone by it had been a hop kiln. From the front door I could see as far as the Black Mountains in Wales, and from the kitchen window I could see Clee Hill in Shropshire. I was thrilled with it.

Just before leaving Moss Hill I had got engaged to Brenda, the sister of George who taught me how to match plough. So I cycled over to her home and told her my good news. I was so excited—the last few hours I had been walking on air—so I chattered away about how we would milk, keep a small pedigree flock of sheep, grow our own food—and when could we get married? Would it be possible to get married and move in before Christmas? We could not be too long, as the first rent was due in March. When I came to leave, however, Brenda came to see me off and walked a short way down the lane with me. She told me she was looking forward to getting married, but she did not want to do so until she was twenty-eight. I couldn't believe it—that was eight years away! I asked her why, but she didn't know—she just knew that she was not ready yet. I pedalled home crestfallen, and was in tears when I arrived there. I rushed off to bed, but soon Jean knocked on the door with a cup of tea and asked if I wanted to talk. I didn't, but I told her anyway. I can't remember what was said, but I felt better. It was nice to know that someone took the trouble to listen.

The next couple of months were hectic: ploughing at Corn Hill, then going down to Buckfield to clean out buildings, get ready to receive cows and get around the fences to see if they were stockproof. I did not do much with the house, just swept it out and made it so that I could make a cup of tea. Jean had a box of sandwiches ready for me to pick up as soon as I finished work at Corn Hill. I would then ride down to Buckfield and work until dark. Once again I noticed that although I became tired, the work did not appear to be hard. I suppose it was because an ambition had been realized that I had not expected, so the adrenalin must have been flowing.

I did not see Brenda during this time, not because I didn't want to but because I had a lot of work to do. I hoped that she would come over to see both the farm and myself, but she didn't. At the end of November I went over to see her, but the age at which she thought she would be ready to marry had been extended from twenty-eight years to thirty. So I broke off the engagement and we parted. A few days later a strange thing happened. Over the previous two years we had had a few photographs taken of ourselves with an old Brownie box camera. I received in the post an envelope with all of them, about twelve in all, but in each one Brenda had been cut out. I threw them in the fire.

Mr Rogers came once or twice to see how things were going. I was pleased to see him and was able to keep him up to date. He had two suggestions to make: first that he would like me to keep working for him as often as I could, with any money so earned deducted from my rent, which was a big load off my mind; secondly that I should not buy any tractor or machinery—I could borrow his any time I wanted. What a friend! Why he should bother with me, with the workload that he had on his shoulders, I did not know, but I was happy that he did. It was also about this time that Mr Downing called at Buckfield to wish me luck. I was so pleased to see him, and this gesture was also very much appreciated. Tom, my old work mate, was the next to call and he offered to come and give me a day or two's work.

ONE DAY just before Christmas I got a fright. A big envelope came through the door with OHMS across the front. It was my National Service call-up papers! There was even a rail warrant to go to Cardington. When I was eighteen I had gone to the enrolment office and told them that I would like to serve in the RAF; I had even filled in all the relevant papers. Some time later, however, I received a letter to say that I was in a reserved occupation and would not be required to do my national service. I now wrote back to the Ministry of Defence pointing out their refusal to have me previously. In reply they said that since I had now changed my job from stockman to farmer I was required to do my two years' service. Two years earlier I had been looking forward to going into the RAF but not now—what would happen to my farm? Mr Rogers again came to the rescue. He took my papers to the National Farmers' Union and they were able to help me to stay on my farm.

The big farmhouse at Corn Hill Cop was a wonderful place to spend Christmas. Jean and Mr Rogers were part of a large family of nine brothers and sisters. One brother had been killed in Crete during the war but the rest were very close, and all of them called at some time over Christmas with their children—there were over twenty of them. It was a happy time and I was pleased to be a part of it.

In those days it was expected that the youngest daughter looked after the parents until they died. This had happened with the Rogers family; Myfanwy, the youngest daughter, had been left to look after their father, who had just recently died. They had lived just outside Ludlow on a smallholding. One of the brothers now wanted the smallholding, so Myfanwy had moved out and was spending six months at a time with each brother or sister; this Christmas she had arrived to stay with Ralph and Jean. As I had come to expect with the Rogers family, she was a lovely, kind person and like everyone who met her, I fell in love with her at first sight. But she was my boss's sister—how could I even think of asking her out? Once again I was out of my depth. If I had stopped to think, I would have run a mile, but as usual I ploughed straight in and asked if I could take her out over Christmas. I think she was too nice to hurt my feelings, because two days later we went up to the woods on Dinmoor Hill and enjoyed a good day out. With all the work I had been doing, it did me good to relax with such a lovely person. The next few weeks were a dream—hard work during the day but coming back to Corn Hill and Myfanwy at night. Two months later I asked her to marry me and we were engaged in March. We got married in May—I was 21 in June. We were to spend thirty-seven happy years together, bringing up a family of two boys and a girl.

Farming during this time was moving very fast. For example, when I went to Moss Hill Farm we had six horses, but by 1950 they had been replaced by two tractors—Tom was now a tractor driver. Chemicals were introduced to kill weeds in growing crops, and the soul-destroying job of thistle spudding was at last redundant. Artificial fertilizers came into use, combine harvesters were now becoming common, and round grain-storage tanks were springing up in the old rickyards. Milking parlours were replacing the old snug cowsheds, but they were cold, draughty, concrete-floored places. In the space of three or four years, haymaking had gone from a horse-and-cart operation with the loose hay being pitched, to almost 90 per cent being bailed, and silage came to the fore.

Combined with the milking parlour, this introduced what was, to my way of thinking, a disastrous combination of slurry, silage effluent and concrete. In short, it was the start of the era of factory farming. Agriculture had never moved so fast: it was incredible. Many of the innovative things were very good, but with such speedy progress a few disastrous blunders were made. One of them was the destruction of the farming 'ladder'. I was able to progress from being a shepherd to a skilled farm worker, to the tenant of a small farm of 32 acres, to the tenant of a larger farm of 220 acres, and later, its owner. And there is nothing special about me; many farm workers did the same. From the 1950s on, however, the small farms were slowly swallowed up by the larger farmers, who, with all the modern aids, could now manage more land. This removed a rung or two of the ladder, making it unusable. Now the best a young man can aim for is to be a manager for a large landowner, which is not quite the same as working for yourself.

As an example of how far and fast things have changed, when I moved back to Yorkshire in 1954 there were sixteen farms in a small parish of 459 people; by 1970 there were two. Fourteen farms and fourteen livings had gone in such a small area. This has been mirrored all over Britain. In my view it is a tragic way for farming to have gone. My son, who is still farming 220 acres, is on an oasis: he is surrounded by one neighbour who farms 6,000 acres, one who farms 3,000 acres and one who farms 2,000 acres. His son Paul wants to follow his father on the farm.

He is ten years old: what chance does he have?

THE STORY OF MY HEART

My Autobiography

Richard Jefferies

with a new Introduction by Jeremy Hooker

The Story of My Heart is not an autobiography in the ordinary sense of the word, as it contains no history of the events of Richard Jefferies' life. It is not concerned with his birth or marriage, his actions or fortunes, but is an outpouring of Jefferies' innermost soul. Like many others, he found himself at odds with the world. He saw the beauty of the land, the grandeur of the sea, the interest of life—above all of human life—but he was not satisfied. He longed for more beauty, a fuller grandeur, a deeper interest. This feeling completely mastered him, and in *The Story of My Heart* he poured out with what strength and what skill he possessed the intensity of his longing.

We are delighted to reprint this classic which has inspired and illuminated thousands of people since its original publication in 1883.

128pp 197 x 133mm ISBN 1 903998 19 0 £6.95 pb

AT HOME ON THE EARTH

A New Selection of the Later Writings of Richard Jefferies

Selected and introduced by Jeremy Hooker

Richard Jefferies (1848-87) has long been recognized as an interpreter of English country life and a natural observer with few if any equals. A 'poet-naturalist' with an ecological vision, he was both mystic and realist, and his writings show the beauty of 'wild England' but also the harsh conditions of labouring life in the Victorian countryside. Jeremy Hooker's new selection represents Jefferies at his best, and reveals his contemporary significance.

176pp 234 x 156mm ISBN 1 903998 02 6 £9.95 pb